AF179807

CFD-Compatible RANS/LES Modeling of Transitional and Separated Flows

Jiakuan Xu · Min Chang · Junqiang Bai

CFD-Compatible RANS/LES Modeling of Transitional and Separated Flows

Springer

Jiakuan Xu
Northwestern Polytechnical University
Xi'an, Shaanxi, China

Min Chang
Northwestern Polytechnical University
Xi'an, Shaanxi, China

Junqiang Bai
Northwestern Polytechnical University
Xi'an, Shaanxi, China

ISBN 978-981-96-6885-4 ISBN 978-981-96-6886-1 (eBook)
https://doi.org/10.1007/978-981-96-6886-1

© The Editor(s) (if applicable) and The Author(s), under exclusive license to Springer Nature Singapore Pte Ltd. 2025

This work is subject to copyright. All rights are solely and exclusively licensed by the Publisher, whether the whole or part of the material is concerned, specifically the rights of translation, reprinting, reuse of illustrations, recitation, broadcasting, reproduction on microfilms or in any other physical way, and transmission or information storage and retrieval, electronic adaptation, computer software, or by similar or dissimilar methodology now known or hereafter developed.
The use of general descriptive names, registered names, trademarks, service marks, etc. in this publication does not imply, even in the absence of a specific statement, that such names are exempt from the relevant protective laws and regulations and therefore free for general use.
The publisher, the authors and the editors are safe to assume that the advice and information in this book are believed to be true and accurate at the date of publication. Neither the publisher nor the authors or the editors give a warranty, expressed or implied, with respect to the material contained herein or for any errors or omissions that may have been made. The publisher remains neutral with regard to jurisdictional claims in published maps and institutional affiliations.

This Springer imprint is published by the registered company Springer Nature Singapore Pte Ltd.
The registered company address is: 152 Beach Road, #21-01/04 Gateway East, Singapore 189721, Singapore

If disposing of this product, please recycle the paper.

Contents

Chapter 1
Introduction

Abstract Boundary layer transition refers to the phenomenon where fluid shifts from laminar to turbulent flow, causing significant changes in wall friction forces, heat flux, and flight mechanical characteristics. Therefore, accurate transition prediction is crucial for research in aerospace, wind engineering, and related fields. This book will provide comprehensive coverage of development methodologies for subsonic/transonic boundary layer transition models, formulation of Very Large Eddy Simulation (VLES) approaches applicable to separated flows and construction techniques for transitional-separated flow hybrid models.

1.1 Research Background and Significance

Over recent years, the continuous advancement of computing power has significantly matured Computational Fluid Dynamics (CFD) methods, which brings about their widespread application in aerospace, automotive engineering, shipbuilding, and turbomachinery industries. In the design of aircraft, wind turbines, and turbomachinery, CFD offers an efficient and cost-effective approach to simulate complex atmospheric conditions encountered during flight; it also stimulates the operational environment of engines at high altitudes. It enables the accurate and comprehensive analysis of aerodynamic characteristics and engine performance at a fraction of the cost, thereby providing crucial insights for design optimization. Historically, the engineering use of CFD, particularly over the past two to three decades, has focused primarily on simulating flow fields for conventional designs such as aircraft, ships, and turbine blades. To meet design requirements, these configurations are typically designed to limit flow separation and maintain smooth flow field gradients. At the same time, turbulence models used in steady-state simulation offer adequate predictive accuracy. Furthermore, as optimization techniques evolve and higher safety and environmental standards are required, researchers are increasingly challenged by more complex issues, such as simulating large-scale flow separation and stall, predicting and controlling noise in various components, and analyzing dynamic loads. Spalart and Bogue [1], along with Sagaut and Deck [2], have pointed out that

© The Author(s), under exclusive license to Springer Nature Singapore Pte Ltd. 2025
J. Xu et al., *CFD-Compatible RANS/LES Modeling of Transitional and Separated Flows*, https://doi.org/10.1007/978-981-96-6886-1_1

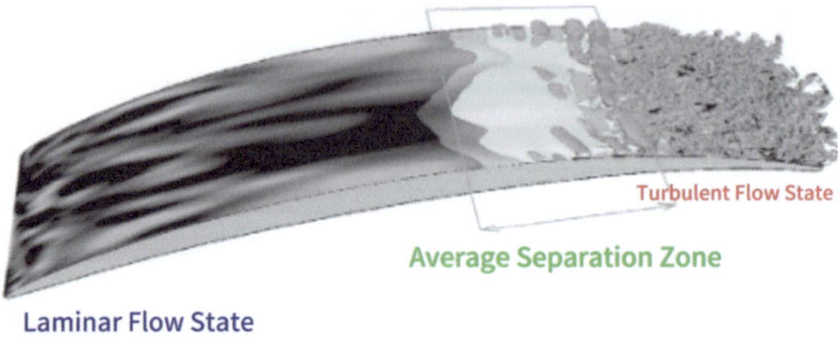

Fig. 1.1 Numerical simulation of laminar separation bubble-induced transition on a compressor blade [4]

addressing these intricate problems requires CFD methods for three-dimensional unsteady flow simulations.

Although turbulence dominates most natural flows, the phenomenon of laminar-to-turbulent transition is a recurring challenge in the design of aircraft, automobiles, and turbomachinery. Laminar flow presents notable aerodynamic benefits, including substantial reductions in skin friction, aerodynamic heating, and noise. For instance, the HondaJet, designed with a natural laminar flow nose, achieves a reduction in drag of approximately 10% compared to a turbulent nose configuration [3]. Nonetheless, the laminar boundary layer is considerably less stable than its turbulent counterpart, which makes it more prone to laminar separation bubbles, particularly under strong adverse pressure gradients. Figure 1.1, based on direct numerical simulation (DNS) by Zakit et al. [4], illustrates the flow field over a compressor blade, where separation on the suction side critically impacts compressor performance, significantly reducing isentropic efficiency.

Furthermore, in complex operational environments, surfaces of aircraft wings or turbomachinery blades can become rough due to factors such as dust accumulation, ice buildup, insect residue, chemical corrosion, or thermal damage (as depicted in Fig. 1.2). Such surface roughness alters the boundary layer transition process, degrading the aerodynamic efficiency of aircraft and turbomachinery, power extraction efficiency in energy systems, and thermal durability of hot-section components. In addition, compared to smooth airfoils, a rough leading edge can lead to a reduction in the maximum lift coefficient, an increase in drag, and a decrease in the zero-lift pitching moment coefficient [5, 6]. Meanwhile, both transition phenomena and surface roughness will influence the development of the boundary layer and perturb downstream separated flows, altering their unsteady characteristics. The accurate simulation and analysis of these roughness-affected transitional flows have thus become critical research gaps in the field. This issue has garnered significant attention, researchers are trying to develop advanced and efficient numerical simulation methods to address these challenges effectively.

Currently, several numerical methods exist for modeling turbulence and its effects on flow simulations, each offering distinct advantages and limitations concerning

(a) Ice Accumulation on the Wing Surface

(b) Corrosion on Wind Turbine Blades (c) Corrosion in Rotating Machinery

Fig. 1.2 Ice accumulation on aircraft wings, corrosion on wind turbine blades, and corrosion in rotating machinery

computational cost and predictive accuracy. Theoretically, precise turbulence simulation is achievable; nonetheless, according to Spalart [7], applying high-precision numerical methods to complex geometries in engineering will require decades of further development. To make numerical simulations feasible within practical computational constraints, modeling assumptions regarding turbulence characteristics are essential. These simplifications introduce uncertainties that inevitably induce predictive discrepancies across models under identical flow conditions. In general, turbulence simulation methods can be broadly classified into three primary approaches: Direct Numerical Simulation (DNS), Large Eddy Simulation (LES), and Reynolds-Averaged Navier–Stokes (RANS) methods.

Turbulence concludes three-dimensional vortices of varying scales. The majority of the turbulent kinetic energy is concentrated in the largest vortices, whose length scales are on the order of the characteristic dimensions of the object or the flow. These large-scale vortices extract energy from mean shear through vortex stretching, hence exhibiting high anisotropy due to mean flow dominance. The energy cascade process, driven by vortex stretching, transfers energy from these larger scales to progressively smaller vortices. DNS represents the most comprehensive approach to resolving unsteady turbulent flows. It involves the direct discretization and numerical solution of the governing equations. Moreover, it requires no any turbulence modeling, and resolves the complete spectrum of turbulence scales in both space and time when properly resolved. Owing to the inherently three-dimensional and unsteady nature of turbulence, DNS addresses a four-dimensional problem. As the Reynolds number increases, the Kolmogorov scales diminish in size relative to the largest scales, amplifying scale separation and thus escalating computational resolution requirements. Besides, the motion of the smallest vortices must be resolved temporally. At the same time, temporal resolution should align with the spatial resolution, which further amplifies computational requirements. In high Reynolds number regimes (10^5–10^8), such as flows with, boundary layers, and free shear layers, the span of spatiotemporal scales becomes significantly exceeds current computational resources' capacity for full-scale resolution. Although DNS is an effective and high-precision method for studying fundamentals of transition and turbulence, as well as control strategies, its application remains impractical for most engineering scenarios due to its prohibitive computational costs at high Reynolds numbers.

LES methods are fundamentally grounded in spatial filtering techniques. In LES, a cutoff scale, or filter width, is defined to segregate the turbulent flow into resolvable and subgrid-scale (SGS) components. According to Kolmogorov's hypothesis, which characterizes the turbulent energy spectrum, small-scale turbulence, at sufficiently high Reynolds numbers, can be approximated as locally isotropic. Consequently, during the spatial filtering process, vortices larger than the cutoff scale are resolved directly, while those smaller than the cutoff scale are modeled using SGS stress models. The effectiveness of LES stems from its capacity to accurately resolve large-scale turbulent structures, which significantly contribute to the transport of mass, momentum, and energy within the flow. Compared to DNS, LES reduces computational expenses by implementing empirical modeling techniques for smaller-scale turbulence. Nevertheless, LES retains the requirements of DNS in terms of three-dimensional spatial resolution and temporal unsteadiness. With higher Reynolds numbers, computational demands will increase sharply. Therefore, while LES provides detailed insights into large-scale turbulence, its application in practical engineering regimes, such as simulating the complete flow field of an aircraft, often incurs substantial computational costs and may deliver excessive detail in less critical regions of the flow field.

As a result of these limitations, traditional methods for predicting high-Reynolds-number separated flows often rely on solving the Unsteady RANS (URANS) equations. This approach can yield relatively accurate results for complex industrial geometries, such as attached flow simulations around aircraft. Nonetheless, it encounters substantial difficulties when addressing large-scale separation flows in simpler geometries. This challenge primarily stems from the strong correlation between the dominant vortex structures in large separation flows and the geometric configuration, which involves significant temporal dynamics. Classical RANS turbulence models are often inadequate in capturing the temporal evolution of these turbulent structures, leading to reduced accuracy in these flow predictions.

The divergence between RANS and LES methods gave rise to the introduction of hybrid RANS/LES methods. The central concept of this approach is to apply the LES method, which resolves small-scale turbulence, to regions outside the boundary layer and within separated flow zones, while relying on the RANS method for turbulence modeling in the near-wall regions. This strategy reduces the grid resolution required near walls and lowers computational costs, while simultaneously resolving turbulent fluctuations in the LES regions to enhance the accuracy of unsteady flow predictions. Given its balance between computational efficiency and predictive precision, hybrid RANS/LES methods have gained increasing adoption in various engineering applications. One of the earliest attempts to simplify near-wall treatment in LES, which laid groundwork for later Wall-Modeled LES (WMLES), was made by Schumann [8] in 1975. He proposed a wall model applied within the LES framework to avoid resolving the smallest scales near the wall. For all that, significant attention to hybrid RANS/LES methods specifically combining RANS and LES regions was not garnered until later contributions from Speziale [9, 10] and Spalart et al. [11] in the 1990s. Their pioneering work marked a transformative step, enabling the application of hybrid RANS/LES methods to industrial problems. These breakthroughs brought hybrid approaches to the forefront of turbulence modeling research, CFD software development, and industrial applications, driving continuous advancements and improvements in hybrid methods.

The hybrid RANS/LES method demonstrates robust predictive capabilities for separated flows, even though most existing hybrid methods are tailored primarily for fully turbulent flows, overlooking the common transition from laminar to turbulent flow in real-world engineering applications. These methods often struggle to accurately capture the impact of the transition process on the evolution of separated flows. In turn, investigating transition phenomena and separated flows particularly under conditions involving surface roughness, and developing a precise and efficient prediction method for flow transitions and separations (especially on rough surfaces) within the hybrid framework, holds substantial theoretical importance and practical significance for engineering applications.

1.2 Research Progress on Hybrid RANS/LES Methods

Over the past two decades, significant progress has been made in developing hybrid RANS/LES methods to enhance the numerical prediction of complex flows in engineering applications. These methods typically employ the RANS model in near-wall regions and the LES mode in areas further from the wall, especially in separated flow zones. The computational cost of such hybrid approaches remains manageable for practical use. In regions of the flow where greater physical detail is required, finer grids and LES are used to capture turbulent vortex interactions more accurately, while in less complex areas, RANS is sufficient to provide satisfactory results. As Chaouat [12] notes, it is possible to integrate RANS and LES within a mathematically unified framework (e.g., via partial averaging) by harmonizing the levels of physical representation. Since both the ensemble-averaged Navier–Stokes equations and the filtered equations can be expressed in similar mathematical forms, RANS and LES can be coupled through different closure models to construct an effective hybrid approach. Based on these coupling strategies, the literature [13, 14] generally divides hybrid methods, into two categories: zonal hybrid methods and global hybrid methods.

The zonal hybrid RANS/LES method employs two distinct turbulence models: a RANS model for near-wall regions and a SGS LES mode for regions away from walls or with significant separation, with a clear interface dividing the two zones. When transitioning across the interface, the turbulence closure method switches from one model to another, often giving rise to discontinuities in the flow solution, particularly in velocity and stress fields. The mismatch in energy spectra between RANS and LES modes, especially since the RANS model typically predicts much higher stress than LES, tends to create discontinuities at the interface. To maintain accurate velocity and stress distributions in the boundary layer, stochastic fluctuations are introduced at the RANS-to-LES interface to compensate for missing turbulent spectral content, thereby improving continuity across the models. Piomalli et al. [15] adopted an artificial source term through a stochastic backscatter model, while Davidson [16], Larsson [17], and others used DNS and LES data to construct synthetic fluctuations. Shur et al. [18] provided a comprehensive review of synthetic turbulence generators designed for zonal hybrid methods at the RANS/LES interface. As well, Temmerman et al. [19] incorporated damping functions at the interface to reduce RANS-predicted stresses, thereby mitigating discontinuities at the interface. For the LES-to-RANS transition, in light of the significant disparity in the turbulent spectra, not all information from the LES region can be smoothly transferred to the RANS region. Accordingly, modifications to the RANS model are required. Zhang and Shih [20] improved the downstream RANS model by using eddy viscosity tensors reconstructed based on upstream LES turbulence statistics, along with a weak equilibrium assumption. This approach enhanced both the prediction accuracy and the stability of the hybrid method.

The global hybrid RANS/LES method is recognized for its ability to maintain a more continuous flow field, as both the RANS and LES modes share a unified equation structure across the entire domain. This approach ensures consistency in the energy spectrum space and maintains a coherent turbulence solving equation throughout the flow field. An early conceptual non-zonal approach, later termed Very Large Eddy Simulation (VLES), was pioneered by Speziale [9, 10]. This method employs a conventional RANS turbulence model to compute the Reynolds stress tensor and then adjusts the ratio of modeled to directly solved turbulence components using a scale control function. Its advantage lies in the theoretical ability to achieve a smooth transition between DNS and RANS accuracy. The mathematical formulation for the modeled stress in the VLES method initiated by Speziale [9, 10] is given by the following equation:

$$\tau_{ij} = \left[1 - \exp\left(-\frac{\beta\Delta}{L_K}\right)\right]^n \tau_{ij}^{(R)} \tag{1.1}$$

In the equation, $\tau_{ij}^{(R)}$ denotes the stress term computed using the Reynolds stress model, Δ represents the computational grid scale, L_K is the Kolmogorov length scale, and β and n are model constants. Building on Speziale's foundational work, researchers such as Zhang et al. [21], Fasel et al. [22, 23], and Israel [24] have advanced the VLES method from various perspectives, with Fasel et al. [22, 23] later formalizing this class of methods as the Flow Simulation Methodology (FSM). It is crucial to clarify that FSM is not a specific model but a collective term for a range of methodologies. These methods bridge RANS models and alternative approaches through the use of stress contribution functions within a unified turbulence solving framework, thereby allowing for adjustable resolution scales. The core principle of FSM involves determining the "physical resolution" by comparing instantaneous local turbulence-related length scales with local filtering length scales. This process enables the appropriate scaling for turbulence modeling and precisely represents unresolved sub-grid stresses. Han et al. [25–28] have developed novel flow resolution functions from an energy perspective and developed new VLES models based on the standard $k-\varepsilon$ model and the Wilcox $k-\omega$ model. These models have been validated across multiple turbulent flow scenarios and have been effectively applied to combustion and heat transfer simulations [29–31], demonstrating both accuracy and reliability. Furthermore, Liu et al. [32] put forward the Partially Resolved Numerical Simulation (PRNS) method, and Zhang et al. [33] proposed the Continuous Scale Switch Equation (CSSE), both of which fall within the FSM framework. Distinct from FSM, another prominent and widely adopted global hybrid RANS/LES method is the Detached Eddy Simulation (DES) method, initiated by Spalart [11] in 1997. This method, based on the one-equation Spalart–Allmaras (SA) turbulence model [34], involves reconstructing the destruction term and introducing a grid scale filter to facilitate the transition between RANS turbulence models and LES subgrid models. In the DES method, the wall distance d_w used in the SA model is replaced with a length scale $\tilde{d}$ that incorporates the grid scale, thereby enhancing the model's capability to handle both near-wall and large-scale turbulent structures:

$$\tilde{d} = \min(d_w, C_{\mathrm{DES}}\Delta), \tag{1.2}$$

In this context, the grid scale Δ is given by $\Delta = \max(\Delta_1, \Delta_2, \Delta_3)$, where Δ_1, Δ_2, Δ_3 are local grid spacings, and the calibration coefficient C_{DES} is set to 0.65 based on the attenuation of uniform turbulence. Strelets [35] further advanced the DES method by applying it to the Menter $k-\omega$ shear stress transport (SST) model. In this extension, a blending length scale l_{DES} was incorporated into the dissipation term of the turbulent kinetic energy transport equation, thereby refining the model's ability to switch between RANS and LES formulations:

$$D_{\mathrm{DES}}^{k} = \frac{\rho k^{\frac{3}{2}}}{l_{\mathrm{DES}}},\ l_{\mathrm{DES}} = \min\left(\frac{k^{\frac{1}{2}}}{\beta^{*}\omega}, C_{\mathrm{DES}}\Delta\right), \tag{1.3}$$

For the SST model, the $k-\omega$ model is used within the boundary layer to maintain accuracy in simulating adverse pressure gradients, while the $k-\varepsilon$ model is applied outside the boundary layer to avoid the non-physical behavior of the SST model near the wall. As a consequence, the model coefficient C_{DES} must account for the effects of both regions. Strelets redefined the model coefficient using the F_1 function from the SST model:

$$C_{\text{DES}} = C_{\text{DES}}^{k-\omega} \cdot F_1 + C_{\text{DES}}^{k-\varepsilon} \cdot (1 - F_1), \tag{1.4}$$

Subsequently, Li and Ren [36] developed a DES method based on the $k - \varepsilon$ model, implementing it in fire simulations. In 2003, Menter et al. [37] first addressed the issue of grid-induced separation in DES methods and introduced the SST-DDES method, which effectively mitigates premature activation of the subgrid model within the boundary layer. Building on this concept, Spalart et al. [38] formulated the SA-DDES method based on the SA model and improved the grid sensitivity of DES methods. Shur et al. [39] launched the IDDES method, which rectified the log-layer mismatch issue inherent in DES methods. Gritskevich et al. [40] recalibrated model parameters to advance the DDES and IDDES methods based on the $k - \omega$ SST model. More recently, Yin et al. [41, 42] introduced an adaptive DDES method that achieved accurate reproduction of critical flow structures compared to DNS data. He et al. [43] developed a dynamic DDES model based on the $k - \omega$ SST model, which can address the log-layer mismatch by dynamically adjusting model coefficients.

1.3 Transition Prediction Methods

On a theoretical basis, DNS can precisely capture the entire transition process from laminar to fully turbulent flow by solving the three-dimensional, unsteady Navier–Stokes equations. To resolve the fine-scale turbulence structures, DNS requires highly refined grids. With advancements in computational technology, DNS has been applied to transition simulations in flat plates and low-pressure turbine blade cascades [4, 44]. The strength of DNS lies in its ability to provide comprehensive transient and averaged data across the entire flow field, which makes DNS databases a valuable tool for developing and testing turbulence models. Nevertheless, in light of its immense computational demands, DNS is not yet suitable for routine engineering applications. LES offers a way to further reduce computational costs and has been successfully used to simulate transitional flows [45–47]. Nonetheless, the transition location is highly sensitive to the choice of the Smagorinsky constant and the type of subgrid model employed. Subsequent developments in dynamic subgrid models [48, 49] have effectively mitigated this issue. However, for most engineering applications, the computational cost of LES remains prohibitively high.

Early transition models relied on empirical relationships to identify transition locations. These models, grounded in extensive experimental data, established correlations between the transition Reynolds number and parameters such as incoming turbulence intensity and local pressure gradients. The transition point was determined by comparing these empirical criteria with the momentum thickness Reynolds number derived from laminar flow solutions. Prominent empirical criteria include Michel's criterion [50], Eppler's criterion [51], and the Abu-Ghanam and Shaw model [52]. Models based on stability theory provide a more physically grounded approach to predicting transition. The e^N method, developed by Smith [53] and van Ingen [54], utilizes linear stability theory and the parallel flow assumption. Arnal [55] and Cebeci [56] extended this method to three-dimensional compressible flows. This semi-empirical approach requires experimental calibration, as the N-factor value at the transition onset is not universal and typically ranges from 7 to 9. The theoretical basis of the e^N method limits its ability to handle with nonlinear phenomena such as bypass transition or roughness-induced transition, and it also faces challenges in predicting separation-induced transition. Bertolotti et al. [57] advanced the e^N method using nonlinear stability theory, though this approach still necessitates boundary layer integration to compute non-local boundary layer variables along streamlines. Implementing the e^N method in modern CFD codes, which employ parallel or unstructured solvers, presents significant challenges. Drela and Giles [58] simplified the e^N method by leveraging linear stability theory and similar velocity profile analyses, establishing a functional relationship between the most unstable amplification factor and the boundary layer shape factor in the streamwise direction. Building on this, Coder and Maughmer [59] presented an amplification factor transport equation and solved it using an approximate envelope method to predict the amplification factor along the boundary layer. This transition model is capable of predicting Tollmien-Schlichting (T-S) wave transition and laminar separation bubble-induced transition. By using local flow information, this model is compatible with modern CFD codes, thereby offering a practical solution for integration into current computational frameworks.

In the transition process, the flow exhibits intermittent behavior, alternating between laminar and turbulent states. This phenomenon is commonly characterized using an intermittency factor. The intermittency factor is defined as [60]:

$$\gamma = \frac{T_t}{T} \tag{1.5}$$

In the equation, T is the total statistical time and T_t is the duration during which the flow exhibits turbulent fluctuations. If $\gamma = 0$, it indicates that no turbulence is present throughout the observation period, thus the flow is entirely laminar. Conversely, if $\gamma = 1$, it implies that the flow is consistently turbulent throughout the entire period. The intermittency factor provides a measure of the proportion of time the flow spends in a turbulent state. Libby [61] was the first to introduce the intermittency factor transport equation, which was later refined by Dopazo [62]. Cho and Chung [63] developed a three-equation $k-\epsilon-\gamma$ transition model tailored for free shear layers. Steelant and Dick [64] proposed a transport equation for the turbulence weighting factor, which incorporates both the intermittency factor and free-stream parameters. Extending the work of these pivotal studies, Suzen et al. [65, 66] initiated an improved intermittency factor equation by incorporating acceleration factors to adjust the critical momentum thickness Reynolds number in pressure gradient regions. Langtry and Menter [67, 68] advanced this approach with the $\gamma - \overline{Re_{\theta t}}$ transition model, which integrates transport equations for both the intermittency factor and the transition momentum thickness Reynolds number. This model estimates the onset of transition based on the turbulence intensity and pressure gradient at the boundary layer edge, allowing for local resolution of all variables and making it compatible with contemporary computational fluid dynamics (CFD) methods and large-scale parallel computing capabilities. Subsequently, Menter et al. [69] launched a single-equation γ transition model, which simplifies computational efforts by solving only one additional transport equation while ensuring that the γ equation adheres to Galilean invariance.

Mayle and collaborators [70–72] introduced the concept of laminar fluctuation kinetic energy and developed a transport equation specifically for this quantity upstream of the transition point. They held the view that transition begins when the level of laminar fluctuation kinetic energy in the boundary layer reaches a critical threshold. Expanding on this approach, Walters et al. [73, 74] localized the laminar fluctuation kinetic energy transport equation and proposed the $k_L - k_T - \omega$ transition model. In this framework, laminar fluctuation kinetic energy (k_L) is modeled within the $k_L - k_T - \omega$ paradigm, while turbulent fluctuation kinetic energy (k_T) accounts for the three-dimensional turbulence fluctuations. The $k_L - k_T - \omega$ model has been validated against numerous test cases, which demonstrates its capability to qualitatively predict bypass transition. Fu and Wang [75–77] introduced laminar fluctuation kinetic energy as an additional transported variable alongside turbulent kinetic energy to develop a $k - \omega - \gamma$ model tailored for supersonic and hypersonic flows. They extended this model to effectively simulate compressible bypass transition as well as laminar separation bubble-induced transition. Subsequently, Juntasaro and Ngiamsoongnirn [78] proposed an enhanced $\gamma - k_L$ model based on the $k - \omega$ SST model, which has shown robust performance in simulating both natural and bypass transitions. Xu et al. [79, 80] advanced the $\gamma - k_L$ model using linear stability theory, achieving notable accuracy in predicting natural transition, laminar separation bubble transition, and crossflow transition.

In addition, Jacobs and Durbin [81] as well as Zaki and Durbin [82] utilized DNS to investigate and understand the mechanisms of bypass transition. Based on the DNS data, Durbin [83] and Ge et al. [84] put forth a novel transport equation for intermittency to predict bypass transition. This RANS-based transition model is tailored for simulating bypass transition and has shown good agreement with both experimental and DNS data. Moreover, Kubacki and Dick [85, 86] developed an algebraic model for predicting transition in attached and separated boundary layers, leveraging the concept of intermittency. This model has demonstrated notable effectiveness in turbomachinery flows. Both Durbin's and Dick's transition models were formulated on the basis of Wilcox's $k - \omega$ turbulence model [87, 88].

1.4 Rough Surfaces Simulation Methods

While most flow studies focus on smooth surfaces, rough surface phenomena are frequently encountered in various practical contexts. Early investigations into roughness effects primarily addressed issues such as the fouling of ship hulls by algae and barnacles, or the impact of roughness on pressure loss in cast pipelines [89]. In natural flows, surface roughness arises from factors like vegetation covering terrain, rocks in rivers, and other geological features. In aerospace applications, surfaces may become slightly roughened due to aging paintwork, or significantly roughened by ice accretion on leading edges of wings. Similarly, deposition, corrosion, erosion, and wear can result in rough surfaces on turbine blades [90]. Over the past century, surface roughness has remained a key focus in fluid mechanics research. While qualitative assessments of the effects of rough surfaces on fluid flow are possible, quantitatively characterizing these effects with physical precision presents a substantial challenge. Describing the roughness itself involves numerous complex parameters, including statistical distribution height, width, shape, and density of roughness elements on a surface. Experimentalists face the challenge of precisely measuring flow parameters without introducing invasive instrumentation, all while ensuring the necessary spatial and temporal resolution. Even capturing the topology of basic roughness elements is difficult, let alone achieving resolved measurements of surrounding flow fields. As a result, numerical simulation methods have emerged as indispensable tools for exploring the influence of surface roughness on fluid dynamics.

There are multiple approaches for predicting flow over rough surfaces. The most detailed computational approach involves directly computing the flow around individual roughness elements. Given current computational limitations, DNS or LES methods are frequently used to explore the flow physics of simple rough configurations, as extensively reviewed by Lu and Liou [91], and Busse and Sandham [92]. For more complex and practical geometries, researchers like Bons et al. [93] have utilized RANS methods to investigate rough turbine blades. Nonetheless, direct numerical simulation still demands considerable computational resources and is restricted to specific application contexts. Hence, researchers have focused on modeling roughness effects within the RANS framework as a more efficient approach.

Corrections for turbulence models to account for wall roughness can be broadly divided into two approaches: the discrete element method (DEM) and the equivalent sand grain roughness method. DEM, derived from Schlichting's [94] foundational work, is applicable to surfaces with fully known roughness geometries, where the roughness elements are arranged in a regular and periodic manner. DEM directly incorporates the effects of roughness into the boundary layer equations. Robertson [95] brought up blockage coefficients into the momentum and energy equations to represent the influence of roughness elements on flow blockage, surface drag, and heat transfer. Finson [96] further advanced this approach, and Coleman et al. [97] offered a more rigorous derivation of the underlying equations. McClain et al. [98, 99]

extended the method to handle more common roughness geometries and distributions. While DEM offers a detailed representation of roughness effects, it necessitates modifications to the governing flow equations. Therefore, it makes its implementation complex and limits its adoption in standard solvers. An alternative, widely used in engineering applications, is the equivalent sand grain roughness method. This concept, experimentally established by Nikuradse [89] and later generalized by Clauser for boundary layers, characterizes surface roughness using an equivalent sand grain height. This height is introduced as a new parameter into turbulence models, enhancing near-wall turbulence to account for the increased drag caused by pressure variations over the rough elements. Different turbulence models have been adapted to incorporate roughness corrections. Durbin et al. [100] modified the two-layer $k - \epsilon$ model to account for rough wall conditions, which enables accurate prediction of velocity profile shifts in the logarithmic layer. Suga et al. announced wall functions to analyze the effects of surface roughness on turbulence and heat transfer. Wilcox brought forward modifications to the wall dissipation rate in the $k - \omega$ model to capture the influence of small-scale roughness. Aupoix et al. [101] extended the SA model [34] to rough-wall flows by developing a correction method based on the relationship between wall eddy viscosity and equivalent sand grain roughness. Hellsten and Laine [102] further modified the switching function F_2 in the $k - \omega$ SST model, ensuring that the function deactivates in the viscous sublayer, thereby allowing Wilcox's roughness correction to be applied within the SST framework. Knopp et al. [103] developed a set of rough wall formulations for both the Wilcox $k - \omega$ and $k - \omega$ SST models, which introduces corrections for both turbulence kinetic energy and dissipation rates at the wall. Their results demonstrated good agreement with experimental data for fully rough surfaces. Aupoix [104] identified discrepancies in Knopp's predictions for velocity shifts in the transitional roughness regime, thereby prompting a recalibration of the wall turbulence kinetic energy and dissipation rate based on the experimental findings of Nikuradse [89] and Colebrook et al. [105].

Developing further on these cornerstone work, researchers have sought to model the effects of surface roughness on the laminar-to-turbulent transition process. One approach involves incorporating geometric parameters of the roughness elements or equivalent roughness height into the criteria for determining the onset of transition. Through comprehensive analysis of experimental data on rough turbine blades, Stripf et al. [106, 107] conducted a novel transition correlation that accounts for both turbulence intensity and roughness topology, including parameters such as shape and density. While this method successfully captures the influence of roughness characteristics, it necessitates the integration of boundary layer thickness and involves non-local variables, which limits its broader applicability. Ge et al. [108] and Wei et al. [109] launched the concept of effective origin displacement, they used equivalent roughness height to modify wall distance and source terms within the transport equations. Comparison with experimental results demonstrated that their model provided accurate predictions of transition onset on rough flat plates and turbine blades, which offers a practical alternative to more computationally expensive methods. An alternative modeling approach introduces a scalar transport equation associated with equivalent roughness height, which is employed to delineate the flow region influenced by surface roughness. This scalar is then incorporated into calibrated relations that modify the transition model's governing criteria, thus accounting for the influence of roughness on transition dynamics. Dassler et al. [110, 111] were the first to implement this methodology in the $\gamma - \overline{Re_{\theta t}}$ transition model developed by Langtry and Menter [67, 68], and they launched a roughness amplification factor transport equation to quantify the intensity of roughness effects. Subsequently, Langel et al. [112, 113] enhanced this approach by modifying the production term within the $\overline{Re_{\theta t}}$ equation, thereby capturing the influence of roughness on transition onset. With the increase in roughness height, the production term diminishes within the boundary layer, leading to earlier transition onset. Yang and Xiao [114] refined the effective length scale in the $k - \omega - \gamma$ model [115], which enables the model to more accurately capture the effect of roughness on the first instability mode. Liu et al. [116], building upon the $k_T - k_L - \omega$ model [74], extended it to account for the influence of roughness on both the first and second instability modes. It results in an improved effective length scale. As surface conditions shift from smooth to rough, the near-wall turbulence length scale increases, which reduces the characteristic timescales of turbulence timescale models. This, in turn, accelerates the growth of unstable models and causes an earlier onset of transition. These advancements in roughness modeling represent a significant step forward in the predictive accuracy of transition processes influenced by surface roughness, providing enhanced reliability for complex flow simulations across a variety of engineering applications.

1.5 Main Contributions and Structure of This Study

Based on the preceding analysis, this study advances VLES method by analyzing and refining it, with a focus on commonly encountered flow transition scenarios in engineering and the effects of surface roughness on transition and flow separation. The study involves modeling and analyzing both transition and roughness effects, ultimately establishing an enhanced VLES method that accounts for these factors. The detailed contributions of each chapter are outlined as follows:

This chapter is the part of introduction. This chapter outlines the research background and reviews the current state of research both domestically and internationally. It provides a focused discussion on the development of turbulence simulation methods, transition prediction techniques, and wall surface roughness modeling approaches.

Chapter 2 is the part of the development and enhancement of the Very Large Eddy Simulation method. This chapter first derives the mathematical modeling foundation of hybrid RANS/LES methods, analyzes the inherent deficiencies of the VLES method based on its construction characteristics, and proposes improvements. The newly developed VLES method is validated using case studies including Taylor–Green vortex, 30P30N multi-segment airfoil, and flow around single and square cylinders. The results demonstrate that the improved method exhibits high simulation accuracy in predicting turbulence dissipation and flow separation.

Chapter 3 introduces the study of bypass transition and separation-induced transition prediction methods. This chapter addresses the prediction of bypass transition and separation-induced transition. A concise and effective transition prediction model is proposed within the RANS framework. The model utilizes localized indicators to describe variations in turbulence intensity and pressure gradients and recalibrates the criteria for bypass transition. For the transition process induced by laminar separation bubbles, local information is used to compute the laminar boundary layer shape factor, and the criteria for separation bubble transition are adjusted based on experimental data to accurately reflect the impact of incoming turbulence intensity on bubble size. The new transition model shows excellent agreement with experimental and DNS data in multiple classic test cases.

Chapter 4 is the part of study on transition prediction methods considering wall roughness effects. Based on the transition model developed in this work, this chapter focuses on modeling the effects of wall roughness. It begins by developing a transport equation for a roughness growth factor, A_r, which is related to the equivalent sand grain height. This equation is used to identify the flow regions affected by surface roughness based on the distribution of A_r in the flow field. Thereafter, the transition criteria of the model are adjusted using functional relationships that incorporate A_r, thereby reflecting the impact of roughness on the transition process. The effectiveness and accuracy of the model are validated through tests on pressure gradient plates, the NREL-S814 airfoil, and high/low-pressure turbines.

Chapter 5 shows the modeling and application of VLES Methods considering transition and roughness effects. This chapter integrates the transition and roughness models with the VLES method. By implementing measures such as limiting the laminar boundary layer solver, protecting free-stream turbulence, and constructing transition influence functions for laminar separation bubbles, the combined transition-roughness model is coupled with the VLES method to analyze separation flows affected by transition and rough surfaces. The model is validated using various test cases, including smooth plate separation flow, rough cylinder wake, and rough compressor blade wake. The predictions of the newly developed VLES method show excellent agreement with experimental results and LES data, which demonstrates the model's strong adaptability and high simulation accuracy.

Chapter 6 introduces the summary and outlook. This chapter summarizes the work and innovations presented in this study. It highlights the contributions made and identifies areas where further improvements are needed. Additionally, it provides an outlook on future research directions that warrant continued investigation.

Appendix A describes the similarity solutions of laminar boundary layers, which is the foundation of establishing local-variable-based transition model.

Chapter 2
The Development and Enhancement of the Very Large Eddy Simulation Method

Abstract This chapter provides a detailed analysis of the inherent limitations in traditional Very Large Eddy Simulation (VLES) methods. To address these issues, a targeted improvement strategy is proposed, incorporating a novel filtering scale function and a boundary layer protection function. These advancements aim to mitigate grid-induced separation and modeled stress depletion phenomena. By enabling accurate delineation between Reynolds-Averaged Navier-Stokes (RANS) and Large Eddy Simulation (LES) regions while reducing grid-scale sensitivity, the enhanced approach significantly improves the VLES method's predictive accuracy for fully turbulent separated flows.

2.1 Introduction

With regard to unsteady-dominated separated flow, the RANS method is hard to accurately predict flow development and lacks the capability to capture intricate flow structures. As a case in point, in flow-structure interaction problems, in order to assess resonance with mechanical structures, it is necessary to obtain the frequency of large coherent motions within the flow, which the RANS method cannot achieve. Because of the high computational cost required by DNS and LES methods for large-scale configurations and high Reynolds number problems, it is difficult to apply them in engineering practice. Hybrid RANS/LES modes aim to integrate the RANS and the LES methods, and leverage the advantages of both while mitigating their shortcomings, enabling their integrated application within a unified framework.

© The Author(s), under exclusive license to Springer Nature Singapore Pte Ltd. 2025
J. Xu et al., *CFD-Compatible RANS/LES Modeling of Transitional and Separated Flows*, https://doi.org/10.1007/978-981-96-6886-1_2

2.2 The Development of the Very Large Eddy Simulation Method

2.2.1 Mathematical Foundation of Hybrid Methods

The zonal hybrid approach computes the RANS and LES methods in different regions of the flow field. This approach includes feedback from LES results to the RANS method, or alternatively treats RANS computation results as boundary conditions for the LES method. While conceptually straightforward, the implementation of the zonal approach is notably intricate. This complexity arises because RANS data represents average flow, whereas LES results encompass large-scale turbulent fluctuations. Resultantly, discrepancies in the flow field information manifest at the interfaces between these zones. Furthermore, the zonal approach needs a priori determination of transition boundaries between the two approaches, which limits its applicability to complex flows with unknown flow field characteristics. In contrast, the seamless hybrid approach circumvents these issues by employing a model capable of representing eddy structures across multiple scales to enable a seamless transition between RANS and LES. This approach utilizes a bridging model that constructs a "bridge" function to connect the RANS and LES methods. The bridging model establishes a filtering cutoff scale, enabling the modeling of turbulent flow structures smaller than the cutoff scale, while directly resolving flow structures larger than the cutoff scale.

The concept of the bridging model is derived from the mathematical similarities between the RANS and LES methods. In the LES method, let $\varphi(\vec{x}, t)$ denote the field variables in turbulent flows. In the RANS method, these variables are subjected to ensemble averaging, which produces the decomposition into steady average components and unsteady fluctuating components, as shown by the following equation.

$$\varphi(\vec{x}, t) = \overline{\varphi}(\vec{x}) + \varphi'(\vec{x}, t) \tag{2.1}$$

Ensemble Average + Pulsating Quantity

In the LES method, it is essential to incorporate a filtering scale to filter the variables. Employing a low-pass filter divides the variables into two distinct components: unsteady large-scale low-frequency fluctuations and high-frequency fluctuations, as shown by the following equation.

$$\varphi(\vec{x}, t) = \tilde{\varphi}(\vec{x}) + \varphi'(\vec{x}, t) \tag{2.2}$$

Filter + High-Frequency Fluctuations

By substituting the above two expressions into the Navier–Stokes equations, it shows that the two methods have mathematical similarities in their momentum equations. In the context of the RANS method, the ensemble-averaged momentum equation is formulated as

$$\frac{\partial u_i}{\partial t} + \frac{\partial (\overline{u}_i \overline{u}_j)}{\partial x_j} = -\frac{\partial p}{\partial x_i} + \frac{\partial}{\partial x_j}\left(\nu \frac{\partial u_i}{\partial x_j}\right) - \frac{\partial \tau_{ij}^{\text{RANS}}}{\partial x_j} \tag{2.3}$$

where τ_{ij}^{RANS} denotes the Reynolds Stress.

In the context of the LES method, the momentum equation after applying the filtering operation is formulated as

$$\frac{\partial \tilde{u}_i}{\partial t} + \frac{\partial (\tilde{u}_i \tilde{u}_j)}{\partial x_j} = -\frac{\partial \tilde{p}}{\partial x_i} + \frac{\partial}{\partial x_j}\left(\nu \frac{\partial \tilde{u}_i}{\partial x_j}\right) - \frac{\partial \tau_{ij}^{\text{LES}}}{\partial x_j} \tag{2.4}$$

where τ_{ij}^{LES} denotes the SGS stress.

From Eqs. (2.3) and (2.4), it is apparent that the RANS method and LES method share a significant structural similarity, distinguished primarily in the physical interpretation of the final term on the right-hand side. The RANS method computes its results by using turbulence modeling, whereas the LES method derives its outcomes through the SGS filtering. Given their substantial similarity, the momentum equations of both methods can be unified and expressed as Eq. (2.5).

$$\frac{\partial \tilde{u}_i}{\partial t} + \frac{\partial (\tilde{u}_i \tilde{u}_j)}{\partial x_j} = -\frac{\partial \tilde{p}}{\partial x_i} + \frac{\partial}{\partial x_j}\left(\nu \frac{\partial \tilde{u}_i}{\partial x_j}\right) - \frac{\partial \tau_{ij}^{\text{model}}}{\partial x_j} \tag{2.5}$$

The mathematical similarities between the two approaches have laid a foundation for the construction of a unified bridging model.

2.2.2 *Speziale's Very Large Eddy Simulation Method*

In 1996, Speziale [9, 10] put forth the first bridging model, known as the VLES method. His pivotal concept involved deriving Reynolds stress from traditional RANS models and subsequently using a scale control function to determine the appropriate local filtering scale. This function adjusts the Reynolds stress magnitude based on local turbulent flow conditions and grid resolution, and allows Reynolds stress to contribute variably across different grid scales. This method ensures that RANS computations are conducted on coarse grids, while on grids with sufficiently high resolution capable of directly resolving Kolmogorov length scales, the computations attain the level of DNS. The mathematical formulation of the modeled stress in Speziale's VLES method is expressed as follows:

$$\tau_{ij}^{\text{model}} = f_{\text{hybrid}}\, \tau_{ij}^{\text{RANS}} \tag{2.6}$$

In the above equation, τ_{ij}^{RANS} represents the Reynolds stress computed by the RANS method, while f_{hybrid} denotes the model's scale control function. Speziale utilizes grid scale and the smallest turbulence length scale (Kolmogorov scale L_K) to construct the scale control function, as its mathematical expression is shown as Eq. (2.7).

$$f_{\text{hybrid}} = \left[1 - \exp\left(-\frac{\beta\Delta}{L_K}\right)\right]^n \tag{2.7}$$

In the equation, β and n are model constants. Typically, n is on the order of $O(1)$, while β is a small quantity on the order of $O(10^{-3})$. During the simulation of a three-dimensional flow field, the grid scale is defined as $\Delta = (\Delta x \cdot \Delta y \cdot \Delta z)^{1/3}$, and the Kolmogorov scale can be represented by Eq. (2.8).

$$L_K = \left(\frac{\nu^3}{\varepsilon}\right)^{\frac{1}{4}} \tag{2.8}$$

In the equation, ν represents the molecular viscosity and ε denotes the turbulence dissipation rate.

According to Eq. (2.7), it is evident that the scale control function ranges between zero and one. In this context, Δ/L_K has the role of defining the model's turbulence resolution. When the grid is refined enough, Δ/L_K tends towards zero, and f_{hybrid} equals zero. At the same time, all turbulence scales are resolved, SGS stress vanishes completely, and the flow field simulation closely approximates DNS. Alternatively, when the grid is very coarse, Δ/L_K tends towards infinity and f_{hybrid} equals one. In this context, all turbulence scales are modeled, and the flow field simulation reverts to the RANS method.

Figure 2.1 illustrates the relationship between the scale control function and the turbulence energy spectrum [13]. In this figure, the upper thick black solid line represents the ideal energy spectrum. The cutoff scale for turbulence dissipation within this spectrum is denoted by the Kolmogorov wavenumber, which is inversely proportional to the Kolmogorov scale L_K. The vertical black dashed line is related to the grid scale, while the lower curve shows the variation of the scale control function with the parameter β. In the context of the scale control function, the parameter n primarily influences the function's slope, while the parameter β dictates the scale below which small-scale structures are modeled. Increasing n leads to a steeper slope of the function, whereas increasing β results in a larger range of turbulence scales being modeled.

In general, the predictive accuracy of this flow field simulation method falls between that of RANS and DNS methods. Hence, Speziale introduced the method he proposed as VLES. Originally, VLES referred to simulating turbulent flows using traditional SGS stress models on coarse grids. Speziale designated this method as VLES to emphasize the model's intent. The objective is to employ a hybrid technique that resolves large-scale eddy structures in turbulent flows while modeling the influences of small-scale eddies. This method is intended to yield more accurate

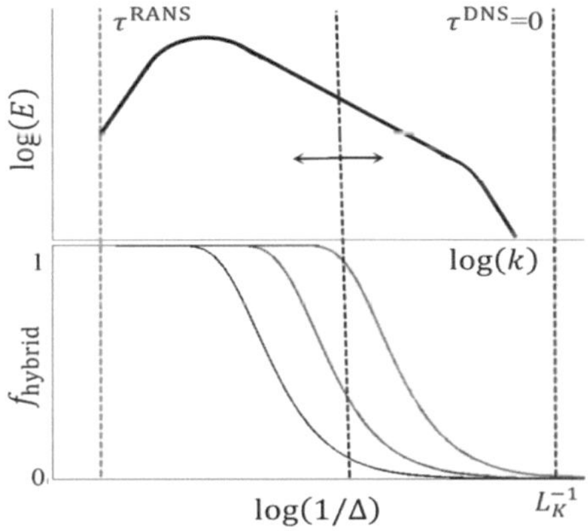

Fig. 2.1 Illustration of the variation of the scale control function

and detailed flow field information. Following Speziale's introduction of the VLES method, Zhang et al. [21] and Fasel et al. [117] implemented the VLES method based on the explicit algebraic stress model (EASM) formulated by Gatski and Speziale [118]. Israel [119] advanced the VLES method from an energy perspective using an enhanced algebraic stress model. Besides, Han and collaborators [25–28] conducted related research employing the standard $k-\varepsilon$ model and the Wilcox $k-\omega$ model. The VLES method brought up by Speziale, as delineated in Eq. (2.6), derives a linear relationship between the turbulent stress and the strain-rate tensor based on the Boussinesq eddy viscosity hypothesis, as follows:

$$\tau_{ij}^{\mathrm{RANS}} \equiv \overline{u_i' u_j'} = \tfrac{2}{3}k\delta_{ij} - 2\nu_t \overline{S}_{ij} \tag{2.9}$$

In the equation, u' denotes the fluctuating velocity in the x direction, ν_t represents the turbulent eddy viscosity, and $k \equiv \overline{u_i' u_i'}/2$ signifies the turbulent kinetic energy. Accordingly, for the eddy viscosity model, the mathematical formulation of the VLES method can be expressed as follows:

$$\tau_{ij}^{\mathrm{model}} = f_{\mathrm{hybrid}}\tau_{ij}^{\mathrm{RANS}} = f_{\mathrm{hybrid}}\left(\tfrac{2}{3}k\delta_{ij} - 2\nu_t\overline{S}_{ij}\right) = \tfrac{2}{3}\, f_{\mathrm{hybrid}}\, k\delta_{ij} - 2\nu_t^{\mathrm{model}}\overline{S}_{ij} \tag{2.10}$$

In the equation, ν_t^{model} denotes the coefficient of SGS eddy viscosity, while ν_t^{RANS} represents the eddy viscosity coefficient obtained through the RANS method. Hsieh [121, 122] adopted this bridging approach in formulating the RANS method. This study learned from a simplified implementation approach inspired by Liu [32] and Batten's [123] treatment of SGS eddy viscosity. Specifically, the equation focuses solely on the treatment of turbulent eddy viscosity, as follows:

$$\nu_t^{\mathrm{hybrid}} = f_{\mathrm{hybrid}}\,\nu_t^{\mathrm{RANS}} \tag{2.11}$$

In the equation,

$$f_{\text{hybrid}} = \left[1 - \exp\left(-\frac{\beta\Delta}{L_K}\right)\right]^n \tag{2.12}$$

and,

$$\Delta = (\Delta x \cdot \Delta y \cdot \Delta z)^{\frac{1}{3}}, L_K = \frac{\nu^{\frac{3}{4}}}{\varepsilon^{\frac{1}{4}}}. \tag{2.13}$$

The model has constant values: $\beta = 0.009,\ n = 1.0$.

2.2.3 The Limitations of the Large Eddy Simulation Method

In theory, the VLES method recommended by Speziale empowers the accuracy of flow field solutions to adapt based on grid density, thus realizing a transition from the RANS method to the DNS method. In contrast to the DES method, which can only achieve simulation accuracy up to the level of LES, the VLES method has a broader range of solution accuracy and greater potential. Nonetheless, in practical computational applications, the VLES method encounters similar drawbacks to other hybrid approaches, along with its own challenges. These issues impede its further application in engineering problems.

Firstly, the initial VLES method proposed by Speziale suffers from excessive attenuation of Reynolds stress. According to the findings of Zhang et al. [21] and Fasel et al. [117], when β in Eq. (2.7) is a quantity of the order of $O\left(10^{-3}\right)$, the computed scale control function is significantly small, even within the boundary layer. Unless the grid is exceedingly sparse, the scale control function cannot approach one, thereby reverting computation to the RANS method. As a subsequent effect, in order to achieve solutions that consistent with physical laws, the grid density near the wall in VLES model calculations must be comparable to that required by LES methods. This makes the VLES method less effective for simulating wall flows at typical engineering grid scales.

Furthermore, Sagaut et al. [14] pointed out that while VLES methods exhibit simulation accuracy levels between DNS and RANS methods, it is difficult to achieve a definitive LES mode when operating within this intermediate accuracy range. Meanwhile, when the Reynolds number is very high, the Kolmogorov scale approaches zero, and the scale control function approaches one, which leads the model behavior to resemble that of a RANS model. At that time, grid refinement has minimal impact on eddy viscosity. In addition, it difficult to transition to LES mode computations regardless of grid refinement efforts.

2.3 Enhancement of the Large Eddy Simulation Method

To address the defects of the initial LES method, this study proposes the following enhancements.

2.3.1 Development of a New Scale Control Function

In the VLES method, the development of the scale control function is of great importance. For flows characterized by a wide range of length scales, the original scale control function contains defects in accurately capturing the resolution of the flow field. This issue is particularly pronounced when the turbulence Reynolds number (defined as the ratio of the largest to the smallest length scales within the flow) varies significantly throughout the entire flow field. Under such conditions, this form of the scale control function may fail to appropriately identify the coarse-grid limit. It suggests that the integral length scale, which represents the largest turbulent structures, plays a critical role in the attenuation of modeled stresses [124]. By drawing an analogy with the DES method and referring to the perspectives of Israel and Fasel [124], Han et al. [27], and Batten et al. [123], the turbulence energy-containing scale, specifically the turbulence integral scale L_i, is introduced as the reference length for scale resolution. Hence, the scale control function f_{VLES} can be reformulated as follows:

$$f_{\mathrm{VLES}} = \left[1 - \exp\left(-\frac{\beta\Delta}{L_i}\right)\right]^n \tag{2.14}$$

The expression for the turbulence integral scale L_i is formulated as:

$$L_i = \frac{k^{\frac{3}{2}}}{\varepsilon} \tag{2.15}$$

In accordance, the model constant β scales as $O(1)$, while n remains the magnitude of $O(1)$. When the grid resolution is sufficiently high, Eq. (2.14) can be expanded in a Taylor series. When $\Delta/L_i \to 0$, $f_{\mathrm{VLES}} \to (\beta\Delta/L_i)^n$, which assumes a form analogous to that analyzed by Sagaut et al. As a result, under this limiting condition, the VLES method approximates traditional LES methods.

Below, the constants β and n in the model are derived, they utilize turbulent kinetic energy k and dissipation rate ε to define the eddy viscosity coefficient.

$$\nu_t^{\mathrm{RANS}} \propto \frac{k^2}{\varepsilon},\ \nu_t \propto f\left(\frac{\Delta}{L_{\mathrm{RANS}}}\right)\frac{k^2}{\varepsilon},\ L_{\mathrm{RANS}} \propto \frac{k^{\frac{3}{2}}}{\varepsilon} \tag{2.16}$$

In the traditional LES method, the SGS eddy viscosity can be expressed as a function of the filter width and the dissipation rate at the resolved scale based on a dimensional analysis [125]:

$$\nu_t \propto \Delta^{\frac{4}{3}} \varepsilon^{\frac{1}{3}} \tag{2.17}$$

Under the equilibrium turbulence assumption, the turbulent dissipation rate remains constant, therefore, the connection between RANS and LES methods can be described as:

$$\frac{\nu_t}{\Delta^{\frac{4}{3}}} = \frac{\nu_t^{\mathrm{RANS}}}{L_{\mathrm{RANS}}^{\frac{4}{3}}} \tag{2.18}$$

Thus, the connection is described as:

$$f\left(\frac{\Delta}{L_{\mathrm{RANS}}}\right) = \frac{\Delta^{\frac{4}{3}}}{L_{\mathrm{RANS}}^{\frac{4}{3}}} \tag{2.19}$$

As a result, it follows that n equals 4/3.

Next, determine the value of β. Under conditions of sufficiently high grid resolution, Δ/L_i approaches zero, and the model resembles LES mode. Under this condition, the SGS eddy viscosity can be expressed as:

$$\nu_t^{\mathrm{sub}} = \left(\frac{\beta\Delta}{L_i}\right)^{\frac{4}{3}} \nu_t^{\mathrm{RANS}} \tag{2.20}$$

In this condition, the model should conform to the traditional "Smagorinsky LES" method, namely:

$$\nu_t^{\mathrm{sub}} = C_s^2 \Delta^2 S, \ S = \sqrt{2 S_{ij} S_{ij}}, \ S_{ij} = \frac{1}{2}\left(\frac{\partial u_i}{\partial x_j} + \frac{\partial u_j}{\partial x_i}\right) \tag{2.21}$$

According to local isotropic turbulence theory, the turbulent kinetic energy dissipation rate at the filtering scale Δ is represented by the following equation [126]:

$$\varepsilon = \frac{1}{\pi^2}\left(\frac{3C_k}{2}\right)^{-\frac{3}{2}} \Delta^2 S^3 \tag{2.22}$$

By combining Eqs. (2.20), (2.21), and (2.22), the solution is obtained as follows:

$$\beta = \left(\frac{\alpha C_s^2}{C_\mu}\right)^{\frac{3}{4}} = 0.459, \ \alpha = \sqrt{\frac{3}{2}\pi^{\frac{4}{3}} C_k} \tag{2.23}$$

The constants are described as follows: $C_s = 0.17$, $C_\mu = 0.09$, $C_k = 0.14$.

In conclusion, the scale control function for the new VLES method is derived as follows:

$$f_{\mathrm{VLES}} = \left[1 - \exp\left(-\frac{0.459\Delta}{L_i}\right)\right]^{\frac{4}{3}} \tag{2.24}$$

2.3.2 Filter Scale and Boundary Layer Protection Function

For global hybrid RANS/LES method, owing to the fact that the transition from RANS model to LES mode is not instantaneous, there exists a "Grey Area" problem. It is the region of transition between RANS and LES modes where the simulation approach is neither RANS nor LES mode. Since the switching function is related to the grid size, refining the grid may cause the hybrid method to switch to the LES mode within the boundary layer. Nevertheless, the grid resolution is not sufficient to solve the velocity fluctuations within the boundary layer, and the Reynolds stresses resolved by the LES mode will not be able to fully replace the Reynolds stresses modeled by the RANS model. This phenomenon is called modeled-stress depletion (MSD). The Reynolds stress depletion causes the surface friction to decrease, which results in the grid-induced separation (GIS) phenomenon. For attached flows, the flow is homogeneous within the boundary layer, and the RANS model is activated. Outside the boundary layer and in the separated flows, the modeled Reynolds stress or turbulent viscosity decreases, and the turbulence instability is released and rapidly enhanced. The simulation filtering scale is changed from turbulence energy-containing scale to grid filtering scale. The time-averaged volume of RANS is combined with the released pulsating volume to form a large-scale filtration resolution and a small-scale modalized volume in the LES mode. The RANS time-averaged quantities are combined with the released pulsation quantities to form the large-scale filter resolution and small-scale modeled quantities in LES mode. Hence, accurate determination of the mode switching location and acceleration of the transition process between RANS and LES modes have been the direction of research efforts.

In the VLES method, the scale control function includes the filter scale, which is closely related to the local grid scale. Furthermore, the filter scale serves as a critical factor in determining the transition point of the flow simulation model. There are multiple definitions of the filter scale Δ, and Table 2.1 enumerates several commonly used definitions.

Breuer et al. [127] highlighted that the filter scale $\Delta_{\max} = \max(\Delta x, \Delta y, \Delta z)$ is not an appropriate length scale for the LES mode region in the DES method.

Table 2.1 Common definitions of filter scale

Δ	Description
$\Delta = (\Delta x \Delta y \Delta z)^{\frac{1}{3}}$	The classical filter scale in LES approach
$\Delta = \max(\Delta x, \Delta y, \Delta z)$	Commonly used filter scale in unified hybrid approaches
$\Delta = 2\max(\Delta x, \Delta y, \Delta z)$	The minimum wavelength that can be captured in any direction by the local grid
$\Delta = \max(\Delta x, \Delta y, \Delta z, \lvert\bar{\mathbf{u}}\rvert \Delta t)$	Filter scale accounting for both spatial and temporal dimensions
$\Delta = \max\left(\Delta x, \Delta y, \Delta z, \lvert\bar{\mathbf{u}}\rvert \Delta t, \sqrt{k}\Delta t\right)$	Filter scale incorporating spatial, temporal, and turbulent fluctuations

To handle this, they utilized the cube root of the cell volume $\left(V^{1/3}\right)$ as an alternative, which improved simulation outcomes. In the IDDES method, different length scale definitions are adopted for flow simulation in the LES mode region. Besides modifying the length scale, more sophisticated blending functions are introduced to ensure the mode operates effectively as part of a wall-modeled large-eddy simulation (WMLES) approach. When turbulent fluctuations are present and the grid resolution is sufficiently high, the blending functions in the IDDES Equations grant the activation of the LES mode within the boundary layer. The adjustment of the length scale mitigates the log-layer mismatch (LLM) in channel flow. In the LES mode region of zonal DES methods, $V^{1/3}$ or Δ_ω is typically used as the length scale. The length scale Δ_ω is determined by the direction of the vorticity ω and the local grid size. Based on prior studies; to expedite the transition between the RANS and LES regions within the boundary layer, this study employs the method proposed by Reddy et al. [128] to define the filter scale:

$$\Delta = f_d V^{\frac{1}{3}} + (1 - f_d)\Delta_{\max} \tag{2.25}$$

In the equation, the blending function f_d is consistent with the coupling function proposed by Spalart et al. [38] in the DDES model. The expression for this function is:

$$f_d = 1 - \tanh\left([C_{d1} r_d]^{C_{d2}}\right), \tag{2.26}$$

$$r_d = \frac{\frac{k}{\omega} + \nu}{\kappa^2 d_w^2 \sqrt{U_{i,j} U_{i,j}}} \tag{2.27}$$

Here, k/ω denotes the turbulent eddy viscosity derived from the RANS model, ν represents the molecular viscosity, κ is the von Karman constant, d_w signifies the distance to the wall, S is the strain rate, and Ω is the vorticity. The model constants are $C_{d1} = 8.0$ and $C_{d2} = 3.0$.

In the VLES method, the scale control function incorporates the grid scale, and also exhibits grid sensitivity issues similar to those observed in the original DES method. When the grid is refined, the scale control function can become too small, prematurely activating the VLES method. Nevertheless, at that time, the grid density may not be sufficient to fully support VLES calculations, which can cause modeled stress depletion. To overcome this issue, a protective function, inspired by the modifications launched by Menter et al. [129] for DES, is introduced into the scale control function. This method constrains the simulation model in the attached boundary layer and mitigates grid-induced separation phenomena. Specifically:

$$f_{\mathrm{VLES}} = \left[1.0 - (1.0 - f_b)\exp\left(-\frac{\beta\Delta}{L_i}\right)\right]^n \tag{2.28}$$

In this context, the protective function $f_b = F_1$ is the blending function from Menter's shear stress transport model [130]. Within the boundary layer, f_b is 1, the scale control function remains at 1, and the flow field is computed by using the RANS method. As the distance from the wall increases, the function rapidly decays to zero, which allows the computation to transition back to the VLES method.

2.3.3 Coupled Turbulence Model

Menter's $k - \omega$ shear stress transport (SST) model [130] is highly regarded for its exceptional accuracy and robustness, which has achieved significant success in various flow simulations. It is frequently employed as a benchmark model for simulating external aerodynamics of aircraft and internal flows within turbomachinery. Accordingly, this study adopts the $k - \omega$ SST model as the foundational turbulence model to construct the VLES hybrid method.

In the $k-\omega$ SST-VLES method, the transport equations for the turbulent quantities are identical to those in the original $k - \omega$ SST model [130], as follows:

$$\frac{\partial(\rho k)}{\partial t} + \frac{\partial(\rho u_i k)}{\partial x_i} = P_k - \beta_k \rho k \omega + \frac{\partial}{\partial x_i}\left[\left(\mu_l + \frac{\mu_t}{\sigma_k}\right)\frac{\partial k}{\partial x_i}\right], \tag{2.29}$$

$$\begin{aligned}\frac{\partial(\rho \omega)}{\partial t} + \frac{\partial(\rho u_i \omega)}{\partial x_i} &= C_\omega P_\omega - \beta_\omega \rho \omega^2 + \frac{\partial}{\partial x_i}\left[\left(\mu_l + \frac{\mu_t}{\sigma_\omega}\right)\frac{\partial \omega}{\partial x_i}\right] \\ &\quad + 2\rho(1 - F_1)\frac{1}{\sigma_{\omega 2}}\frac{1}{\omega}\frac{\partial k}{\partial x_i}\frac{\partial \omega}{\partial x_i},\end{aligned} \tag{2.30}$$

The values of the model constants are consistent with those in Menter's $k - \omega$ SST model [130].

The hybrid approach embeds a scale control function when determining the eddy viscosity, as follows:

$$\mu_t = f_{\mathrm{VLES}} \cdot \mu_{t,\mathrm{RANS}} = f_{\mathrm{VLES}} \cdot \min\left[\frac{\rho k}{\omega}, \frac{a_1 \rho k}{\Omega_2}\right]. \tag{2.31}$$

The turbulent integral length scale L_i is computed based on turbulent kinetic energy k and the specific dissipation rate ω, formulated as follows:

$$L_i = \frac{k^{\frac{1}{2}}}{\beta_1^* \omega} \tag{2.32}$$

In which, the parameter β_1^* is specified as 0.09.

2.4 Case Validation

2.4.1 Flow Field Solver

The computational solver utilized in this study is TeAM (Technology for Aeronautic Multidiscipline), developed by our research group. TeAM features second-order spatial and temporal accuracy and incorporates multigrid and grid sequencing techniques to expedite convergence. For handling convective and pressure terms, the solver adopts an upwind Roe scheme [131] and implements the MUSCL (Monotone Upstream-centered Scheme for Conservation Laws) format [132] for high-order accuracy interpolation across grid interfaces. Stress and heat flux terms are computed by virtue of using central differencing approaches. Time integration is managed via diagonalization [133] using the approximate factorization approach [134]. In unsteady simulations, a dual time-stepping approach [135] is deployed to ensure second-order temporal accuracy through steady-state sub-iterations. The validation of this solver can be referenced in the literature [136, 137], and is not reiterated here.

Further down, the advanced VLES method is put to work to perform numerical simulations of cases in terms of Taylor–Green vortex flow, 30P30N three-element airfoil flow, flow around a single cylinder, and flow around a square cylinder. These simulations are rigorously compared and analyzed against experimental data to substantiate the model's effectiveness and accuracy.

2.4.2 Taylor–Green Vortex

The Taylor–Green vortex (TGV) was brought up by Taylor and Green [138] in 1937 as a canonical flow model for the study of isotropic turbulence. In 1983, Brachet et al. [139] conducted rigorous DNS of TGV, and found out the temporal evolution of dissipation rates across various Reynolds numbers. In 1991, Brachet [140] utilized finer grids, and it produced computational results consistent with earlier findings. By reason of its authority, many researchers established their DNS data as a benchmark for comparative analyses in the field. TGV has since been widely adopted as a standard case for validating sub-grid modeling techniques, calibrating high-order accuracy schemes, and advancing numerical methodologies [141–144]. TGV entails initializing a periodic cubic domain with an initial flow field resembling laminar periodic vortices. Over time, the flow undergoes a transition to turbulence as the periodic vortex structure destabilizes and generates smaller-scale vortices, which contributes to continuous energy dissipation. TGV remains pivotal for exploring flow transition phenomena, cascade processes, and turbulent dissipation mechanisms in fluid dynamics research.

In this case study, the computational domain is defined as $0.0 \leq (x, y, z) \leq 2\pi L$, forming a cube with a side length of $2\pi L$. Periodic boundary conditions are imposed in all three spatial directions. The initial flow field is specified by Eq. (2.33).

$$\rho = \rho_0 \tag{2.33}$$

$$u = V_0 \sin\left(\frac{x}{L}\right) \cos\left(\frac{y}{L}\right) \cos\left(\frac{Z}{L}\right) \tag{2.34}$$

$$v = -V_0 \cos\left(\frac{x}{L}\right) \sin\left(\frac{y}{L}\right) \cos\left(\frac{Z}{L}\right) \tag{2.35}$$

$$w = 0 \tag{2.36}$$

$$p = p_0 + \frac{\rho_0 V_0^2}{16}\left(\cos\left(\frac{2x}{L}\right) + \cos\left(\frac{2y}{L}\right)\right)\left(\cos\left(\frac{2z}{L}\right) + 2\right) \tag{2.37}$$

In the equation, V_0 denotes the initial reference velocity, the reference length of the cube L is 1.0 m, the reference density at the initial moment ρ_0 is given in kg/m^3, and the reference pressure at the initial moment p_0 is 101,325 Pa.

The Reynolds number of the flow is determined by the formula $\mathrm{Re} = \frac{\rho_0 V_0 L}{\mu}$. Assuming the fluid is an ideal gas, the specific heat ratio γ is 1.4, and the Prandtl number Pr is 0.71. To end to that, the initial reference Mach number $Ma_0 = V_0/a_0 = 0.10$, where a_0 is the speed of sound corresponding to the reference temperature T_0. The reference temperature is given by $T_0 = \frac{p_0}{R\rho_0} = 288.5$ K. Assuming the initial temperature field is uniform, the initial density field is calculated using the equation $\rho = \frac{p}{RT_0}$.

At the initial time ($t = 0$), the iso-surface distributions of the velocity components along the X-axis, u/a_0, and along the Y-axis, v/a_0, within the cube are shown in Fig. 2.2. The figure shows the iso-surfaces for $u/a_0 = \pm\ 0.05$ and $v/a_0 = \pm 0.05$. It is apparent that the velocity directions on adjacent iso-surfaces are opposite. Thus, at $t = 0$, the relative motion within the flow field induces the formation of periodic large-scale vortex structures. The characteristic convective time, t_c, is defined as $t_c = L/V_0$, using the reference length L and the initial reference velocity V_0. The simulation for this case study extends until $t = 10t_c$, and a comparative analysis of the results is conducted over this period.

Based on the temporal evolution of the flow, various characteristic quantities can be derived from the flow field. First of all, the time-dependent mean volumetric kinetic energy within the cube is defined as follows:

$$E_k - \frac{1}{\rho_0 V} \int\limits_V \frac{1}{2} \rho \boldsymbol{u} \cdot \boldsymbol{u} dV \tag{2.38}$$

In the equation, V represents the volume of the computational domain. After that, the temporal variation of the kinetic energy dissipation rate, $\epsilon(E_k)$, within the cube can be computed as follows:

$$\epsilon(E_k) = -\frac{dE_k}{dt} \tag{2.39}$$

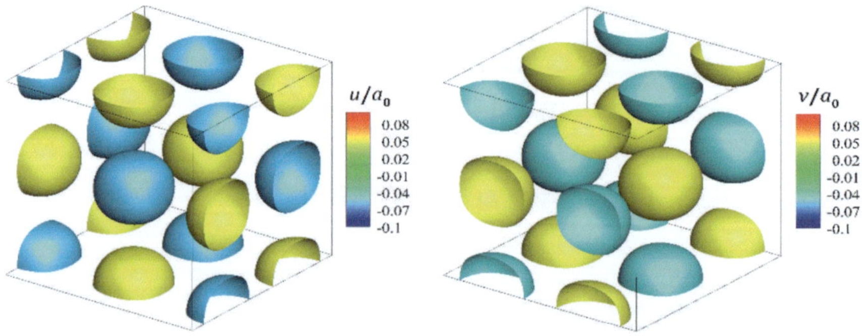

(a) $u/a_0 = \pm 0.05$ (b) $v/a_0 = \pm 0.05$

Fig. 2.2 Iso-surface distribution of transient flow field velocity at initial time

Table 2.2 Status of taylor–green vortex computation

Grid resolution	Time step length (t/t_c)		
	Re = 400	Re = 800	Re = 1600
64^3	0.003	0.002	0.001
128^3	0.003	0.002	0.001

In the numerical simulations, the grid resolution and dimensionless time step are provided in Table 2.2. This case study simulates the flow at three different Reynolds numbers: Re = 400, Re = 800, Re = 1600. Two grid configurations are deployed for the simulations, each with an identical number of grid points in all three directions: 64 points (64^3) and 128 points (128^3).

Compared to the Speziale-VLES method, this study has developed the predictive capabilities of the SST-VLES method and the SST-DDES method [37] for the kinetic energy dissipation rate of the Taylor–Green vortex, results are illustrated in Fig. 2.3. The simulation adopts a grid resolution of 128^3 and a Reynolds number of Re = 1600. The results are benchmarked against the DNS data from Brachet [140] Fig. 2.3 reveals that, up until $t = 4t_c$, the kinetic energy dissipation rates predicted by all three approaches are nearly identical and closely match the DNS data. After that, the predictions begin to diverge: the SST-DDES method overestimates the dissipation rate, while the Speziale-VLES method underestimates it. Regarding the time at which the kinetic energy dissipation rate reaches its peak, the SST-VLES method's predictions are in close agreement with the DNS data, the Speziale-VLES method predicts a delayed peak, and the SST-DDES method predicts an earlier peak. These results show that the SST-VLES method, developed based on turbulent integral scales and equilibrium turbulence theory, has superior predictive capability for turbulent dissipation compared to the Speziale-VLES and SST-DDES methods. In this case study, the simulations were performed on a desktop computer equipped with an Intel Xeon E5-2630v4 processor operating at a base frequency of 2.20 GHz. The computations were conducted on a single block grid by using a single core for all

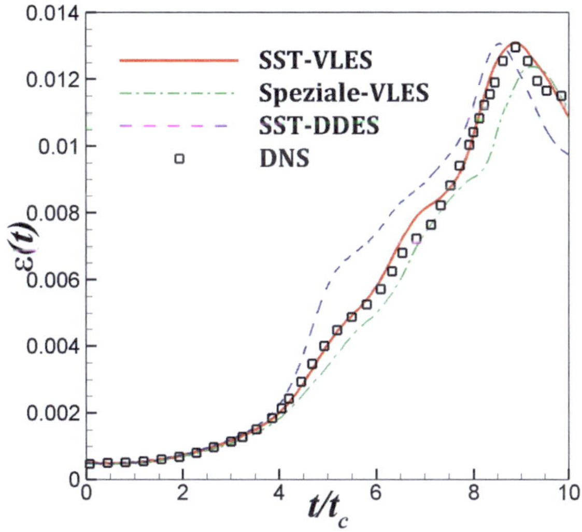

Fig. 2.3 Comparison of predicted kinetic energy dissipation rates for the Taylor–green vortex using different simulation approaches

Table 2.3 Computation times for Taylor–green vortex using different approaches

Method	Cmputation time	Difference (%)
Speziale-VLES	27 h 45 min	–
SST-VLES	28 h 1 min	0.96
SST-DDES	28 h 8 min	1.38

three approaches, which simulates up to 10 t_c. The computation times for the three methods are presented in Table 2.3, showing the SST-VLES method's computational time is comparable to that of the other two methods.

Under different Reynolds numbers and grid resolutions, the temporal evolution of the kinetic energy dissipation rate predicted by the SST-VLES method is shown in Fig. 2.4. The DNS results from Brachet [140] are introduced as a reference. In this case study, it is observed that for all three Reynolds numbers, the kinetic energy dissipation rate initially increases and then decreases over time. In terms of flows with Reynolds numbers of 800 and 1600, the maximum energy dissipation rate (corresponding to the smallest turbulent structures) appears prominently around $t = 9t_c$. With a grid resolution of 64^3, the SST-VLES method's predictions align well with the DNS data during the early stages of flow development. Nevertheless, as time progresses, the SST-VLES results gradually deviate from the DNS data. And the deviation occurs earlier as the Reynolds number increases. When the grid resolution is increased to 128^3, the SST-VLES results maintain good agreement with the DNS data throughout the entire simulation period.

When the grid resolution is 128^3 and the Reynolds number is 1600, the temporal evolution of the flow within the cube predicted by the SST-VLES method is depicted in the figures from Figs. 2.5, 2.6, 2.7, 2.8 and 2.9. In these figures, vortex structures in the flow field use the Q-criterion. The Q-criterion, put forth by Hunt et al. [145]

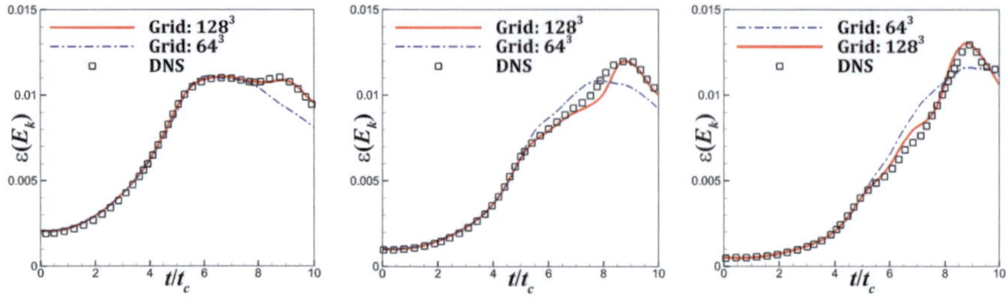

(a) Re=400 (b) Re=800 (c) Re=1600

Fig. 2.4 Evolution of kinetic energy dissipation rates for the Taylor–green vortex predicted by the SST-VLES method at different reynolds numbers and grid resolutions

in 1988, is based on the principle that fluid deformation can be described by the velocity gradient tensor, which can be decomposed into a symmetric part, the strain rate tensor S, and an antisymmetric part, the rotation tensor Ω. Hunt et al. suggested that if the contribution of the rotation tensor to fluid deformation surpasses that of the strain rate tensor, a vortex can be considered to exist in that region. Thus, the Q-criterion is defined as follows:

$$Q = \frac{1}{2}\left(\|\Omega\|^2 - \|S\|^2\right) \tag{2.40}$$

In this equation, $\| \|^2$ denotes the 2-norm of the tensor. The symmetric tensor is defined as $S_{ij} = \frac{1}{2}\left(\frac{\partial u_i}{\partial x_j} + \frac{\partial u_i}{\partial x_i}\right)$, and the antisymmetric tensor is defined as $\Omega_{ij} = \frac{1}{2}\left(\frac{\partial u_i}{\partial x_j} - \frac{\partial u_j}{\partial x_i}\right)$, where $i, j = 13$. The second invariant of the velocity gradient tensor is independent of coordinate selection. As such, this criterion complies with the principles of generalized Galilean invariance. When the Q value is positive, it signifies that the pressure at the core is lower than that in the surrounding regions, analogous to

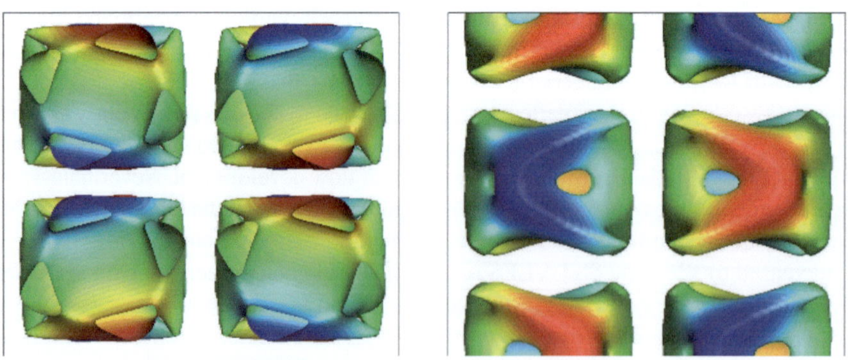

(a) XY Plane View (b) XZ Plane View

Fig. 2.5 Iso-surface distribution of the transient flow field at $t = 2t_c$ with $Q(L/a_0)^2 = 0.01$

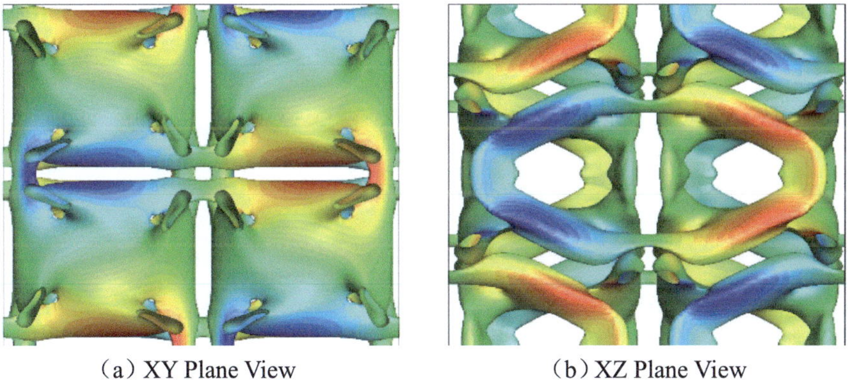

Fig. 2.6 Iso-surface distribution of the transient flow field at $t = 4t_c$ with $Q(L/a_0)^2 = 0.01$

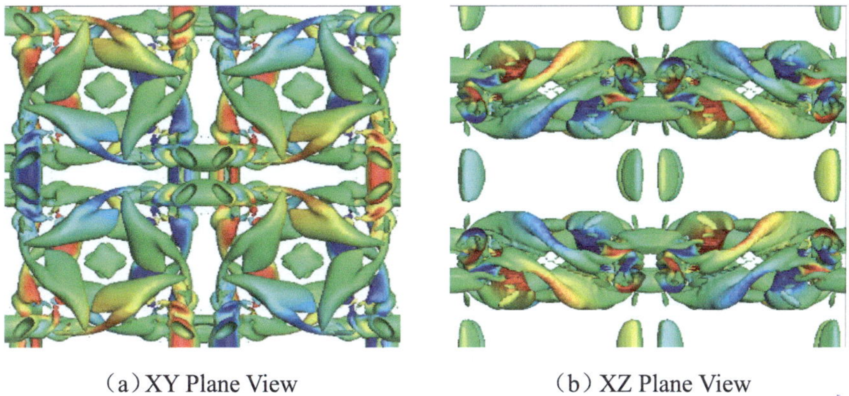

Fig. 2.7 Iso-surface distribution of the transient flow field at $t = 6t_c$ with $Q(L/a_0)^2 = 0.01$

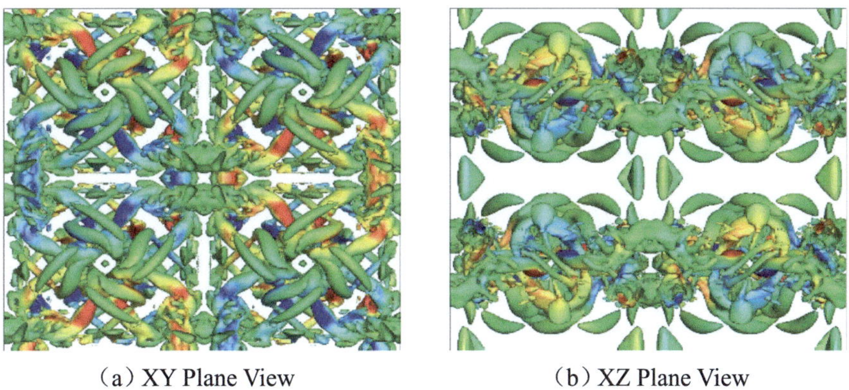

Fig. 2.8 Iso-surface distribution of the transient flow field at $t = 8t_c$ with $Q(L/a_0)^2 = 0.01$

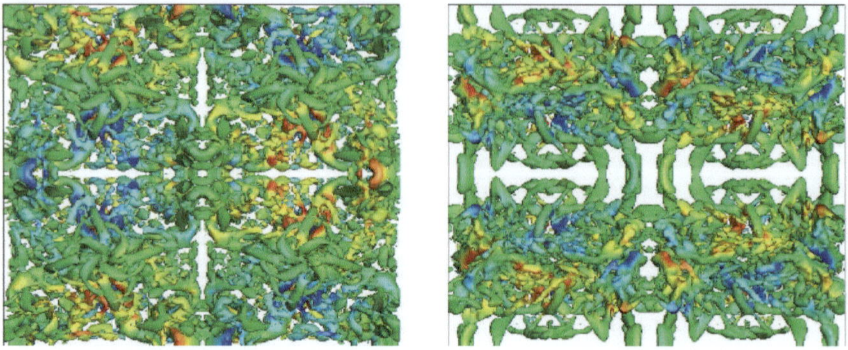

(a) XY Plane View (b) XZ Plane View

Fig. 2.9 Iso-surface distribution of the transient flow field at $t = 10t_c$ with $Q(L/a_0)^2 = 0.01$

the pressure distribution around a vortex core. Accordingly, this metric can effectively capture and represent vortex structures within the flow field.

The transient flow field visualizations depict the iso-surface distribution of $Q(L/a_0)^2 = 0.01$ at various time instances within the computational domain, with local velocity utilized for coloring. The energy dissipation rate in the flow field is primarily governed by small-scale vortices. In Fig. 2.4c, after the onset of flow, the energy dissipation rate progressively increases. It indicates that the interactions between vortices lead to the stretching and deformation of large-scale vortices, which in turn generate small-scale flow structures. The proliferation of small-scale vortices consequently elevates the energy dissipation rate. This phenomenon is consistent with the transient flow field evolution observed in figures from Figs. 2.5, 2.6, 2.7 and 2.8. In these visualizations, as the flow field evolves, large-scale flow structures gradually transition into small-scale flows, leading to increasingly intricate flow structures. When the energy dissipation rate reaches its maximum, the small-scale flow structures are at their most abundant. Thereafter, as these small-scale structures dissipate, the energy dissipation rate decreases, as illustrated in Fig. 2.9 and Fig. 2.4c.

The simulation results of Taylor–Green vortex evolution at three distinct Reynolds numbers indicate that the VLES method developed in this study accurately captures turbulent structures and their dissipation processes. Even under elevated Reynolds number regimes, judicious selection of temporal discretization and sufficient grid resolution ensures that VLES simulations closely match DNS benchmarks.

2.4.3 *Flow Around 30P30N Multi-element Airfoil*

The 30P30N multi-element airfoil configuration represents a benchmark two-dimensional high-lift device case from NASA Langley's 1993 "High-Lift Workshop/ CFD Challenge" [146]. The geometric configuration of the 30P30N multi-element

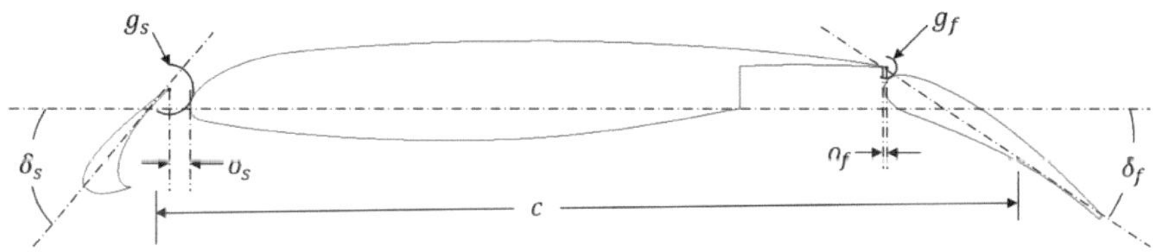

Fig. 2.10 Geometric configuration of the 30P30N multi-element airfoil system

Table 2.4 Geometric parameters of the 30P30N multi-element airfoil configuration

Retracted chord length	c	0.457 m
Slat deflection angle	δ_s	30°
Flap deflection angle	δ_f	30°
Slat gap width	g_s	2.95% c
Flap gap width	g_f	1.27% c
Slat overlap distance	o_s	− 2.5% c
Flap overlap distance	o_f	0.25% c

airfoil is depicted in Fig. 2.10, with its geometric parameters comprehensively detailed in Table 2.4.

Accurately simulating the transition location is particularly challenging due to the impact of incoming flow turbulence intensity and pressure gradients. The experiment was conducted in the Low Turbulence Pressure Tunnel (LTPT) at NASA Langley Research Center. The experimental results include measurements of force, moment, pressure, and surface friction distributions, as well as velocity profiles and transition locations on the upper surfaces of the slat and flap, and both the upper and lower surfaces of the main wing.

Following that, to support noise prediction efforts [147], experimental studies were conducted on the 30P30N three-element airfoil at the Japan Aerospace Exploration Agency (JAXA) [148] and the aeroacoustics wind tunnel at Florida State University (FSU) [149, 150]. The wind tunnel test conditions were standardized, featuring an incoming flow Mach number of 0.17 and a Reynolds number based on the reference chord length c of 1.71×10^6. In addition to measuring forces, moments, and pressure distributions, the experiments encompassed assessments of acoustic sources and power spectral densities at various monitoring points. Stereoscopic Particle Image Velocimetry (SPIV) was utilized at selected angles of attack to examine flow characteristics within the slat cove and the development of shear layers, and provided essential data to support numerical methods for noise prediction.

In this case study, the numerical simulation is conducted with an incoming flow Mach number of 0.17 and an angle of attack of 5.5°, as well as a Reynolds number based on the chord length $\mathrm{Re}_c = 1.71 \times 10^6$. The computational domain is configured as a circle with a radius of 30 c in the $x-y$ plane, which extends 0.1 c in the z-direction, and features periodic boundary conditions in the spanwise direction. Details of the grid setup can be found in Table 2.5, which illustrates both the overall and local mesh configurations as depicted in Fig. 2.11. Accurate simulation of turbulent structures near the wall, separation vortices in the slat cove, and separated flow near the flap trailing edge necessitates appropriate time step settings. In this unsteady simulation, the time step is set to $\Delta t = 3.0 \times 10^{-7}$ s, corresponding to a dimensionless time step of $\Delta t U_{\text{inf}}/c = 3.7 \times 10^{-5}$. Initially, the simulations employ the (URANS method. After that, the SST-VLES method utilizes the URANS results as initial conditions. After the flow field has sufficiently developed (approximately ($TU_{\text{inf}}/c = 0.5$)), time-averaged flow field results are computed, and continue for approximately 3.0 chord-based time units ($TU_{\text{inf}}/c = 0.5$).

Figure 2.12 illustrates the distribution of Q iso-surfaces predicted by the SST-VLES method in the transient flow field, and uses local flow velocity for coloring. From Fig. 2.12b, it is evident that the shear layer originates from the slat tip, undergoes instability generating turbulent eddies that impinge on the slat cove walls, and surrounds the large separation region within the cove. The wingtip vortex from the

Table 2.5 Grid parameters for the 30P30N multi-element airfoil

Total grid count	1.6×10^7
Spanwise grid points for slat	381
Spanwise grid points for main wing	533
Spanwise grid points for flap	305
Height of the first near-wall grid layer	$1.0 \times 10^{-5}c$
Number of grid points in the streamwise direction	61

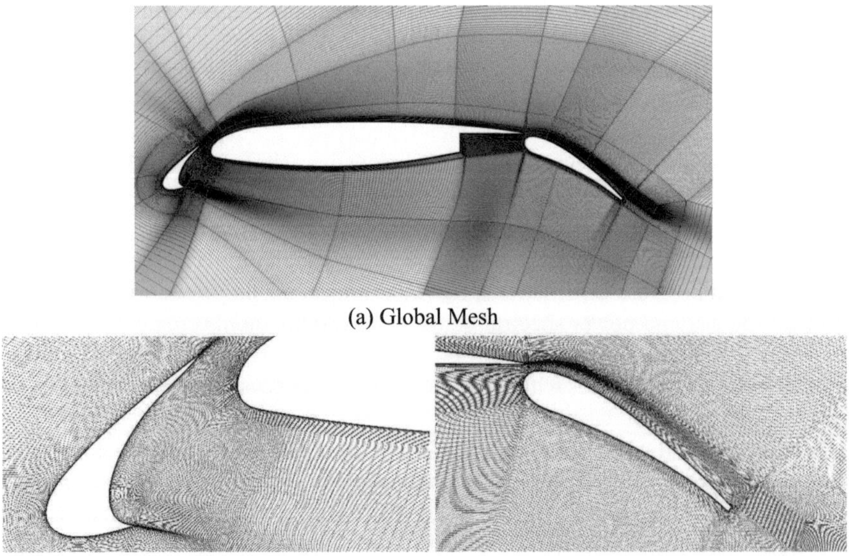

(a) Global Mesh

(b) Mesh around the Slat

(c) Mesh around the Flap

Fig. 2.11 Global and local mesh distribution of the 30P30N multi-element airfoil

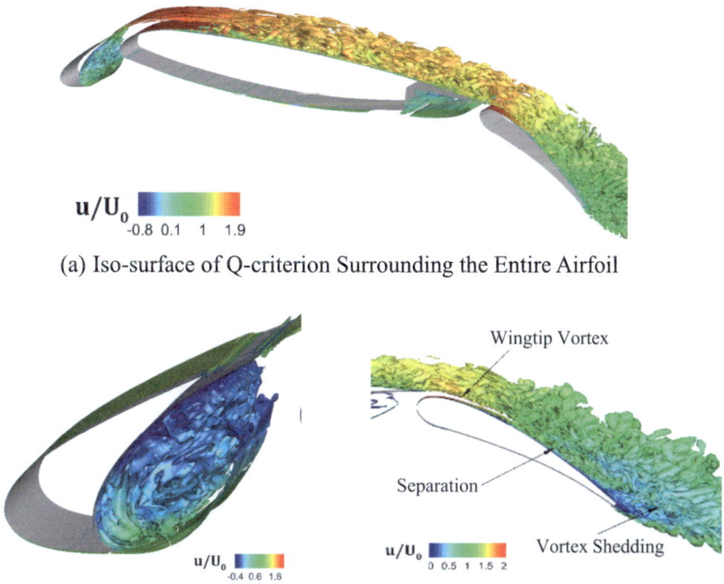

(a) Iso-surface of Q-criterion Surrounding the Entire Airfoil

(b) Iso-surface of the Q-criterion in the Vicinity of the Slat

(c) Iso-surface of the Q-criterion in the Vicinity of the Flap

Fig. 2.12 Distribution of Q-criterion Iso-surfaces surrounding the 30P30N multi-element airfoil predicted using the SST-VLES method

main wing significantly influences the flow over the flap upper surface, as observed in Fig. 2.12, where boundary layer separation occurs at the rear portion of the flap upper surface and forms trailing edge vortices downstream of the flap.

Figure 2.13 shows the mean streamlines in the slat cove, derived from experimental SPIV results [149] and numerical simulations using the Speziale-VLES method, SST-VLES method, and SST-DDES method [37]. Under the influence of the incoming flow, the shear flow beneath the slat impacts the inner surface of the cove, and forms a large separation vortex within the cove. The predictions from the SST-VLES method align closely with the experimental data, and capture the secondary separation flow near the cove walls. In contrast, the separation vortices predicted by the Speziale-VLES and SST-DDES methods are slightly larger along the X-axis and fail to capture the secondary separation flow near the walls.

The distribution of the mean pressure coefficient for the 30P30N airfoil derived from numerical simulations is illustrated in Fig. 2.14. To provide a comprehensive comparison, experimental data from JAXA [148] and FSU [150] are incorporated. Pascioni and Cattafesta [150] have indicated that the experimental conditions at an angle of attack of 10° in the FSU wind tunnel correspond to those observed at an angle of attack of 5.5° in the JAXA wind tunnel. Accordingly, the figure presents the

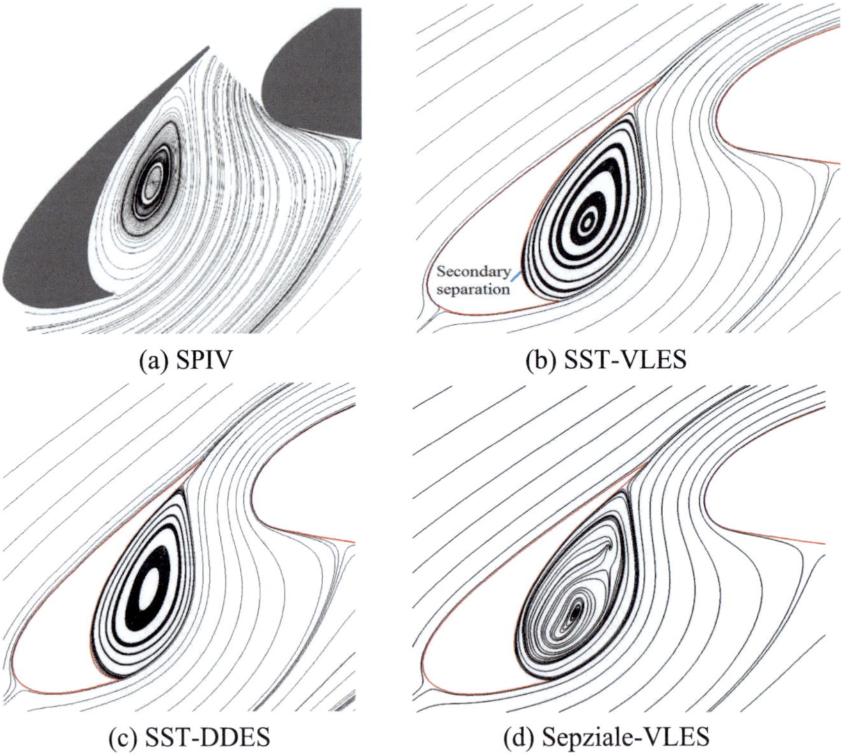

Fig. 2.13 Mean streamlines in the slat cavity of the 30P30N multi-element airfoil

results from the FSU wind tunnel at a 10° angle of attack, alongside the experimental findings from the JAXA wind tunnel at both angles of attack of 5.5° and 6.0°.

The comparative analysis reveals that the SST-VLES method demonstrates good agreement with experimental data in the overall distribution of average pressure coefficients. It precisely predicts pressure peaks on the main wing and flap, as well as accurately simulates boundary layer separation near the flap trailing edge. In contrast, the Speziale-VLES method exhibits lower pressure predictions on the suction side of the flap, shifts the location of flow separation on the flap forward, and overestimates the separation region on the main wing. The SST-DDES method observes that the boundary layer separation near the flap trailing edge but underestimates both the pressure distribution on the flap's suction side and the peak pressures on the main wing, in comparison to experimental data and SST-VLES results. Similar trends were observed in the numerical investigation conducted by Terracol et al. [151].

Near the slat, the distribution of the average velocity magnitude $|U| = \sqrt{U^2 + V^2}$ is depicted in Fig. 2.15. This figure presents results from SPIV [149], along with numerical simulations using the Speziale-VLES, SST-VLES, and SST-DDES methods [37]. The SST-VLES method closely matches the experimental velocity

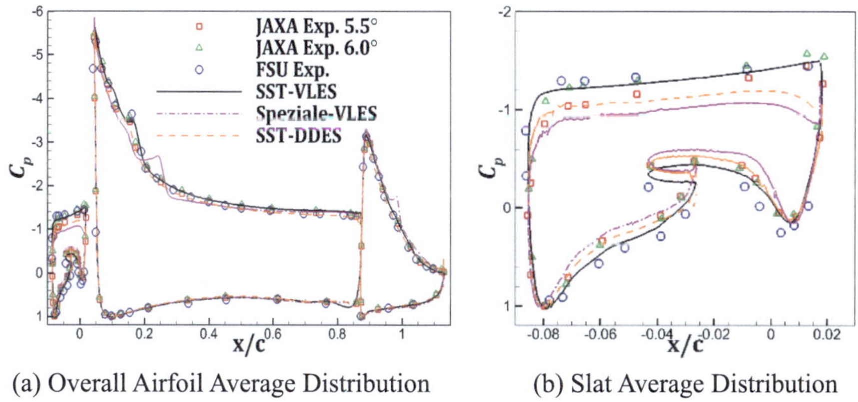

Fig. 2.14 Distribution of average pressure coefficients for the 30P30N multi-element airfoil

distribution. In contrast, the Speziale-VLES and SST-DDES methods simulate a larger stagnation region within the slat cavity compared to the SST-VLES results, consistent with observations in the streamline plots. Figure 2.15a identifies seven measurement lines (L1–L7) positioned within the slat cavity, where flow parameters are monitored experimentally. All seven lines traverse the shear layer within the slat cavity, which allows for a comparative analysis of the numerical methods' accuracy in simulating separation vortices within this region.

The comparison of average velocity distributions along these seven measurement lines is illustrated in Fig. 2.16. In the figure, the regions where velocity increases sharply denote the positions of the separation vortex shear layers. From L1 to L7, both experimental and numerical methods indicate a progressive enlargement of these regions, which reflects the increasing thickness of the separation vortex shear layers. Notably, the Speziale-VLES method shows substantial deviation from the experimental results. For the SST-VLES and SST-DDES methods, the predicted average velocity within the separation zone slightly exceeds the experimental data at L1, L2, and L3. Nonetheless, at other measurement locations, the predicted velocities both inside and outside the separation zones closely match the experimental data. Examining the positions of the separation vortex shear layers predicted by the two numerical methods, it is evident that the shear layer location calculated by the SST-VLES method is in close agreement with the experimental results. Conversely, the shear layer predicted by the SST-DDES method is situated further from the cavity wall and exhibits a greater thickness compared to the experimental results.

The distribution of the mean spanwise vorticity $\Omega_z c/U_{\text{inf}}$ along the measurement lines is presented in Fig. 2.17. This figure includes only the experimental data and the results from the SST-VLES method. On the L1 line, the peak spanwise vorticity predicted by the SST-VLES method is notably higher than the experimental value. All the same, from L2 to L7, the numerical simulation results closely match the experimental data.

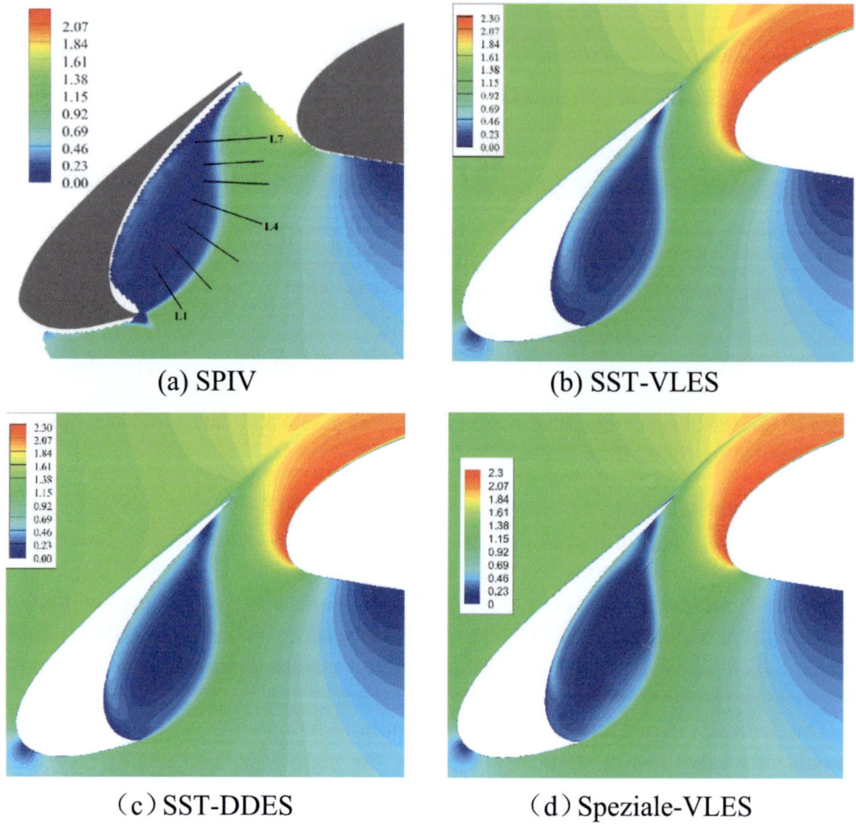

Fig. 2.15 Mean velocity distribution in the slat cavity of the 30P30N multi-element airfoil

The results of this case study demonstrate that the advanced VLES method can predict boundary layer separation phenomena in airfoil flow accurately and offers high simulation accuracy for the separated shear layer.

2.4.4 Flow Around a Single Cylinder

Flow around a cylinder is a classical problem in fluid mechanics, it encompasses complex flow phenomena and intricate physical mechanisms. This scenario includes boundary layer separation and detachment, shear layer instability, wake distribution and variation, and their mutual interactions. Despite extensive research, the fundamental understanding of cylinder flow remains incomplete, rendering it a persistent challenge. Periodic vortex shedding induces changes in the pressure distribution on the cylinder surface, which leads to periodic variations in forces acting in both

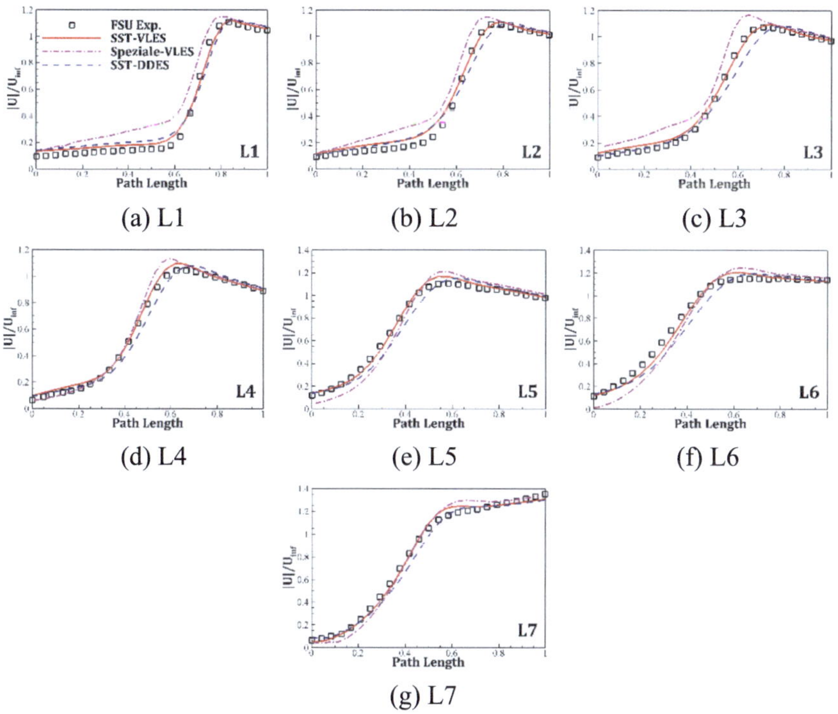

Fig. 2.16 Mean velocity distribution along seven lines through the slat cove shear layer

the streamwise and transverse directions. If the shedding frequency approaches the cylinder's natural frequency, resonance may occur, potentially compromising the structural integrity. This issue is of significant practical importance across various fields, including aerospace, mechanical engineering, hydraulics, and environmental science. In validating hybrid RANS/LES methods, flow around a cylinder remains a classic benchmark case. This study employs simulations of cylinder flow to evaluate the predictive capabilities of the advanced VLES method for boundary layers and large-scale separation phenomena.

In the wind tunnel experimental center [152], the cylinder has a diameter $D = 0.04445$ m, and the freestream velocity is $U_0 = 56$ m/s, which corresponds to a Mach number of 0.166. The Reynolds number based on the cylinder diameter is $Re_D = 1.66 \times 10^5$. In this case study, the computational domain is circular with a radius of $30D$ and a spanwise length of $3D$. An O-type grid is employed for the computational mesh. To accurately simulate the separated flow in the wake of the cylinder, the grid is refined in the wake region. Considering the fully developed flow in the spanwise direction, the cylinder's length is $L = 3D$. The grid resolution consists of 297 points in the circumferential direction, 297 points in the normal direction, and 81 points in the spanwise direction, totaling approximately 7.14 million grid cells. The first grid layer near the wall has a height of $1 \times 10^{-4}L$.

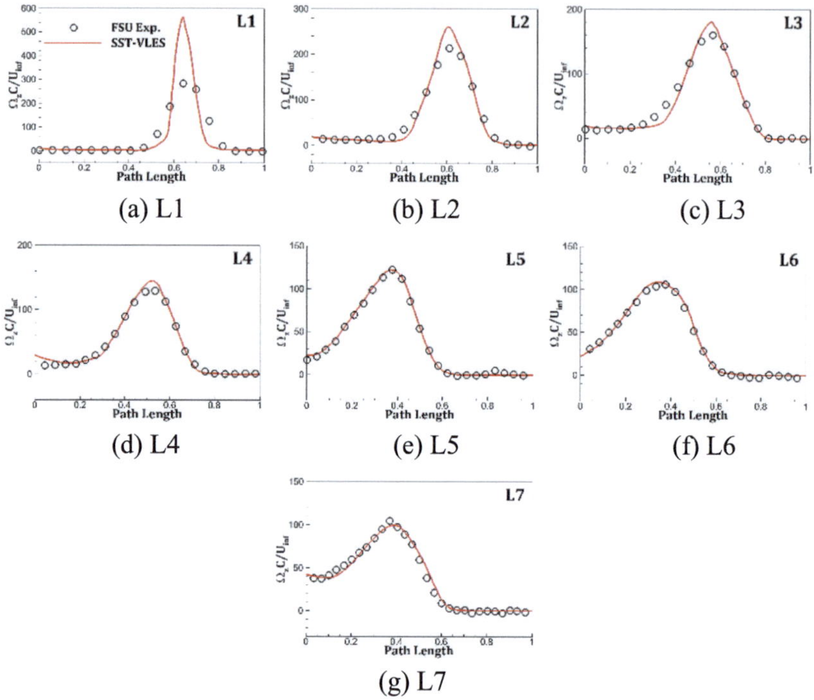

Fig. 2.17 Distribution of mean streamwise vorticity along seven lines across the shear layer in the slat cove

In the far-field of the computational domain, non-reflecting boundaries are implemented. The cylinder surface is characterized as a non-slip wall, and periodic boundary conditions are applied in the spanwise direction on both sides. Figure 2.18 shows the spatial grid topology of the cylinder's computational domain and the grid distribution in its vicinity.

During the numerical simulation of the flow field, the non-steady time step is set to $\Delta t = 0.003D/U_0$. Initially, the flow field is computed by means of using the URANS method. Thereafter, the SST-VLES method utilizes the URANS results as the initial condition. The SST-VLES computation continues until the flow field reaches full development, around $120D/U_0$. Following this phase, statistical averaging of flow variables commences and persists for approximately $600D/U_0$.

In the transient flow field predicted by the SST-VLES method, Fig. 2.19 illustrates the distribution of the scale control function. In the far-field and attached boundary layers, the scale control function f_{VLES} equals 1, which indicates the use of RANS model for simulation in these regions. All the same, within the separated boundary layer and in the wake of the cylinder, f_{VLES} is less than 1, signaling the activation of LES mode. This facilitates the capture and analysis of unsteady turbulent fluctuations and vortex structures.

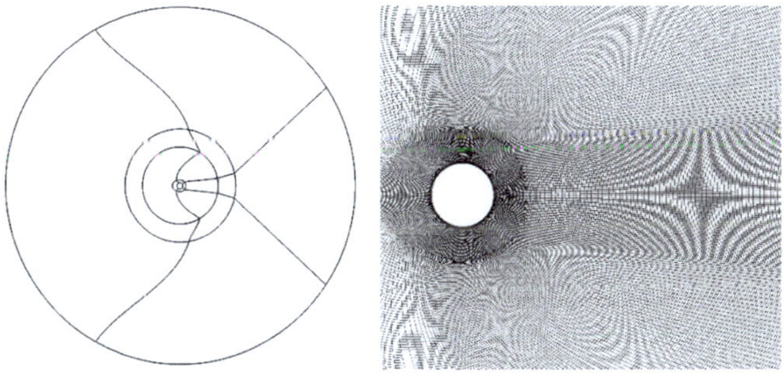

(a) Spatial Grid Configuration (b) Grid Distribution around the Cylinder

Fig. 2.18 Spatial grid topology of the computational domain and grid distribution around the cylinder

Fig. 2.19 Distribution of the transient scale control function f_{VLES} in the cylinder flow field simulated with the SST-VLES method

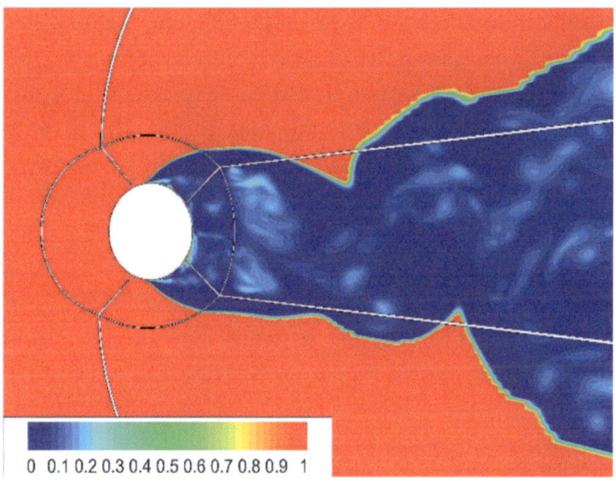

Figure 2.20a depicts the distribution of $Q = 1.5U_0^2/D^2$ iso-surfaces in the transient flow field, and shades it by local streamwise velocity. This visualization aids in identifying coherent flow structures or vortex formations within the field. The LES method effectively captures small-scale turbulent structures present in the flow. In Fig. 2.20b, the distribution of streamwise vorticity $\omega_z D/U_0$ is depicted. It reveals that, at this instance, the cylinder induces separation shear layers of varying lengths on its upper and lower sides. The instability of these shear layers at different locations contributes to the unsteady motion observed in the cylinder wake.

The distribution of mean pressure coefficient (C_p) on the cylinder surface is represented in Fig. 2.21a. This figure compares results obtained by using the SST-VLES method with those from Vatsa and Lockard [153], who employed the MDDES method. The SST-VLES method predicts a C_p distribution on the cylinder surface that closely resembles experimental data. The method precisely predicts peak C_p values

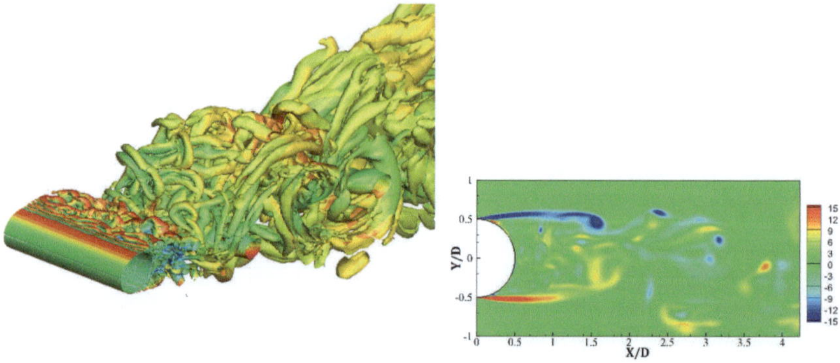

(a) $Q = 1.5U_0^2/D^2$ (b) Transient Streamwise Vorticity $\omega_z D/U_0$

Fig. 2.20 Transient flow field around a cylinder predicted by the SST-VLES method

comparable to those from MDDES but slightly lower than experimental values. While C_p values in the pressure plateau region predicted by SST-VLES are slightly smaller than experimental results, the predicted location of boundary layer separation aligns well with experimental data. In Fig. 2.21b, the root-mean-square (RMS) of the pressure coefficient on the cylinder surface is shown. The RMS values predicted by the SST-VLES method closely match experimental data, with values in the plateau region larger than those predicted by MDDES.

Figure 2.21a pictures the mean streamwise velocity distribution along the centerline downstream of the cylinder. The SST-VLES method simulates a substantial separation region on the leeward side of the cylinder, with predicted separation zone dimensions and velocity distribution closely matching experimental findings.

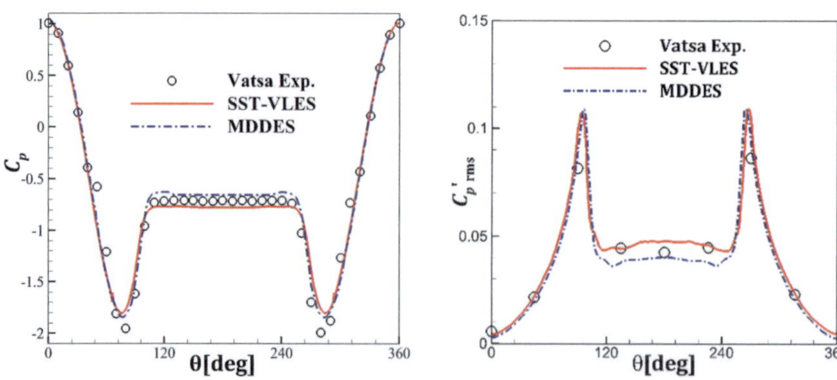

(a) Average Distribution of Cylinder Surface Cp (b) Root-mean-square of Cylinder Surface Pressure Coefficient

Fig. 2.21 Distribution of average pressure coefficient (Cp) and its root-mean-square (RMS) on the cylinder surface

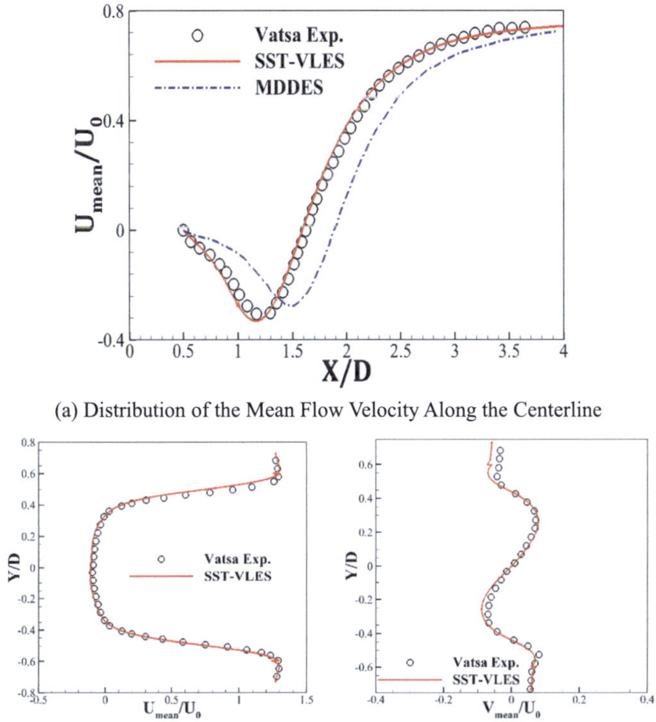

(a) Distribution of the Mean Flow Velocity Along the Centerline

(b) Mean Streamwise Velocity Distribution at X/D

(c) Mean Vertical Velocity Distribution at X/D

Fig. 2.22 Mean velocity distribution in the wake of a cylinder

In contrast, the MDDES method predicts a larger separation zone size with a smaller maximum recirculation velocity. At $X/D = 0.75$, the profiles of average streamwise velocity and vertical velocity are extracted and illustrated in Fig. 2.22b and 2.22c, respectively. The SST-VLES simulation results exhibit excellent agreement with experimental data, and the method demonstrates reliable prediction of the vertical extent of the separation region.

Figure 2.23 displays the mean streamline distribution downstream of the cylinder. Both experimental data and numerical simulation results reveal a significant wake region forming behind the cylinder, with experimental separation ending around $X/D = 1.6$. The SST-VLES method dependably predicts the separation characteristics and length of the wake region, which closely aligns with experimental observations.

Figure 2.24 shows the distribution of mean streamwise vorticity downstream of the cylinder. It reveals that extensive separation regions on the leeward side of the cylinder generate separation shear layers extending downstream from both upper and lower sides of the cylinder. The results from the MDDES method and experimental

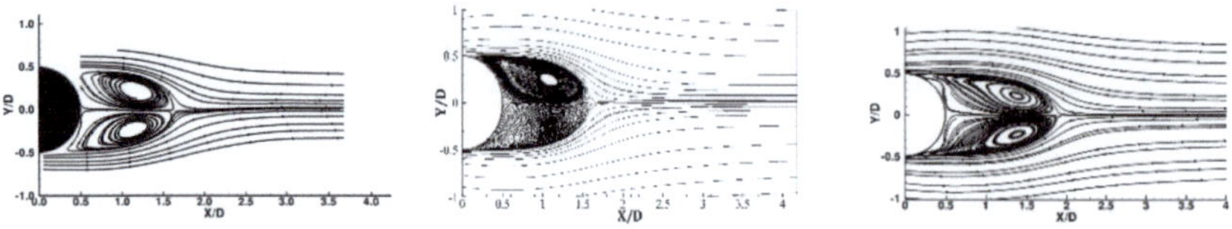

(a) Experimental Data (b) SST-VLES (c) MDDES

Fig. 2.23 Mean streamline distribution in the wake of a cylinder

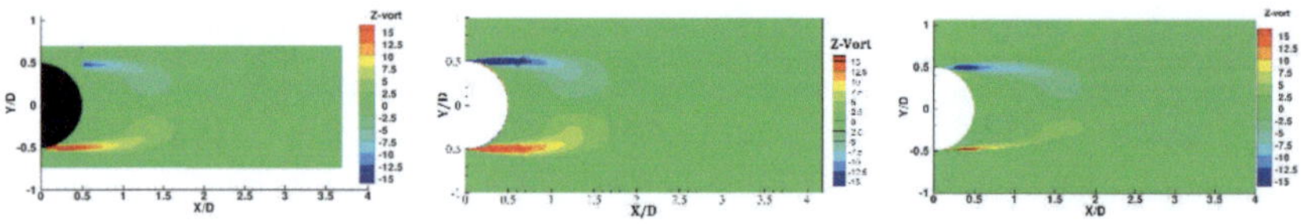

(a) Experimental Data (b) SST-VLES (c) MDDES

Fig. 2.24 Mean spanwise vorticity distribution in the wake of a cylinder

data show that the instability location of these separation shear layers appears further downstream compared to the results from those of the SST-VLES method. This corresponds to the MDDES method's prediction of a longer separation zone in the cylinder wake.

The mean streamwise and vertical velocity distributions downstream of the cylinder are illustrated in Figs. 2.25 and 2.26. As Vatsa and Lockard [153] did not provide the vertical velocity distribution, only the streamwise velocity distribution predicted by the MDDES method is included for comparison. In Fig. 2.25, the flow fields obtained from the three methods all form a "jujube core"-shaped recirculation zone downstream of the cylinder.

Compared to the SST-VLES and experimental results, the MDDES method forecasts that the maximum recirculation velocity is located further from the cylinder. In Fig. 2.26, the average vertical velocity distribution obtained using the SST-VLES method closely matches the experimental data.

Based on the comparison of cylindrical flow results, the VLES method developed in this study correctly captures the separated shear layer in the wake and effectively models the processes of vortex formation, development, and shedding.

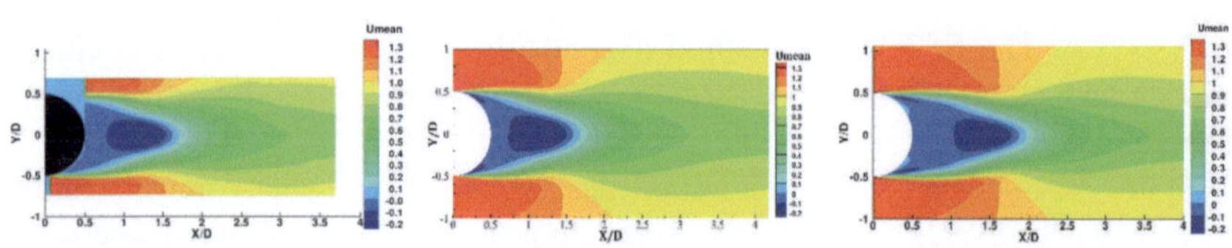

(a) Experimental Data (b) SST-VLES (c) MDDES

Fig. 2.25 Mean streamwise velocity distribution in the wake of a cylinder

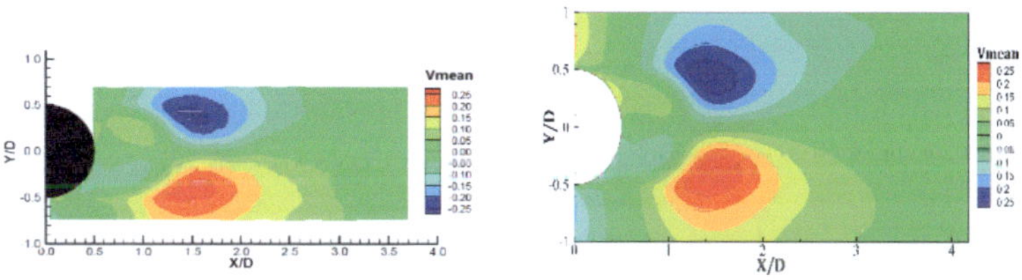

(a) Experimental Data (b) SST-VLES

Fig. 2.26 Mean vertical velocity distribution in the wake of a cylinder

2.4.5 Flow Around a Square Cylinder

Extensive research has been conducted on flow around cylindrical structures, and predominant researches focus on circular cylinders. Oppositely, studies concerning flow around square cylinders are relatively sparse, despite the similarity in flow topology near the wake region. The mechanisms of flow separation on these cylinders, however, differ markedly. For circular cylinders, separation occurs owing to the adverse pressure gradient on the leeward side, with the separation point fluctuating along the cylinder surface due to vortex shedding. Inversely, in the case of square cylinders, the separation point remains fixed at the corners of the windward face owing to the abrupt geometric changes. The distinctive flow characteristics around square cylinders are significant for practical applications in heat and mass transfer processes, such as the design of combustion chambers where the analysis of flow around flame stabilizers is crucial for optimizing performance. Davis and Moore [154] conducted numerical simulations to explore the effects of aspect ratio and angle of attack on forces and wake structures around square cylinders at two Reynolds numbers. Lyn and Rodi [155] performed experimental studies at higher Reynolds numbers, they concentrated on the shear layers formed at the windward angle of attack rather than the wake region. Okajima [156] investigated the relationship between the Strouhal number and Reynolds number for square cylinders with varying aspect ratios and Reynolds numbers. In these studies, the blockage ratio was maintained below 7.5%, and it can minimize the impact of channel walls on the wake. Davis et al. [157] conducted both experimental and numerical investigations of flow around square cylinders with a blockage ratio of 25% situated at the channel center. Key parameters included the wind tunnel blockage ratio and the aspect ratio of the square cylinder, with results indicating that the drag coefficient increased with Reynolds number. Nakagawa et al. [158] undertook experimental investigations on square cylinders with different aspect ratios under similar conditions, with a 20% blockage ratio. They analyzed the pulsating flow structures by decomposing them into coherent and incoherent components and employed phase-averaging and time-averaging techniques to quantify their contributions to the flow field.

The flow domain in this study is configured similarly to the experimental setup of Nakagawa et al. [158], as illustrated in Fig. 2.27. The square cylinder is positioned

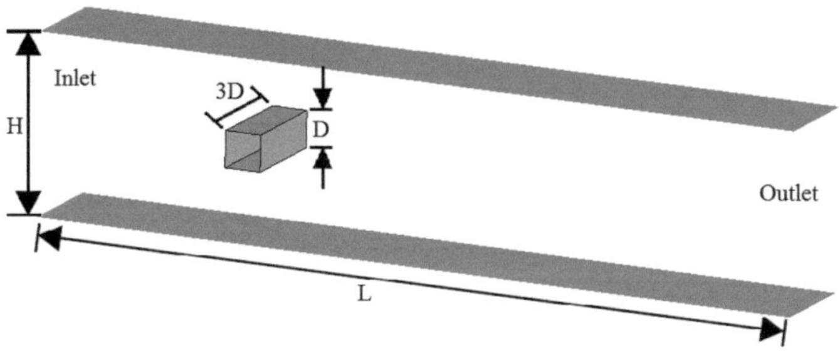

Fig. 2.27 Computational domain for simulating flow around a square cylinder

centrally within the channel, with no-slip wall conditions applied to the upper and lower channel walls, and periodic boundary conditions imposed in the spanwise direction. In this flow configuration, the development of the Kármán vortex street downstream of the square cylinder is significantly influenced by the channel walls, and the flow structures near these walls are also affected by the vortex street.

All variables are non-dimensionalized utilizing the mean inlet velocity (U_m) and the side length of the square cylinder (D). Two grid resolutions, coarse and fine, were generated for the computations. The height of the first layer of grids near the wall in the computational domain is 0.005D. The coarse grid, Grid 1, consists of 1.12 million cells, with grid points in the streamwise, normal, and spanwise directions being 209 × 121 × 49, respectively. The fine grid, Grid 2, consists of 3.25 million cells, with grid points in the streamwise, normal, and spanwise directions being 345 × 145 × 69, respectively. Figure 2.28 illustrates the distribution of these two grid sets in the X–Y plane, with the origin of the X-axis located on the leeward face of the square cylinder.

In this case study, the Reynolds number based on the side length of the square cylinder and the inlet velocity is $Re_D = 3000$. The blockage ratio is 20%, and the spanwise dimension of the flow domain is $3D$. The SST-VLES method and the SST-DDES method [37] are employed to simulate the flow around the square cylinder. The predictions from these hybrid methods are compared against LES data [159] and the experimental results from Nakagawa et al. [158]. In the subsequent discussion, the results from the SST-VLES and SST-DDES methods are respectively labeled as "VLES" and "DDES". In the numerical simulations, once the flow is fully developed, time-averaging and spanwise-averaging are performed to obtain the mean flow field. Figure 2.29 illustrates the distribution of streamwise velocity and normal turbulent fluctuations along the centerline downstream of the square cylinder for different grids and methods. In Fig. 2.29a, the streamwise velocities predicted by the VLES method on both coarse and fine grids align well with the experimental values and LES data. In contrast, the DDES method underestimates the maximum recirculation velocity and the size of the recirculation zone. Additionally, the centerline velocity predicted by the DDES method significantly exceeds the experimental values and LES data as the distance from the square cylinder increases. In Fig. 2.29b, the normal turbulent

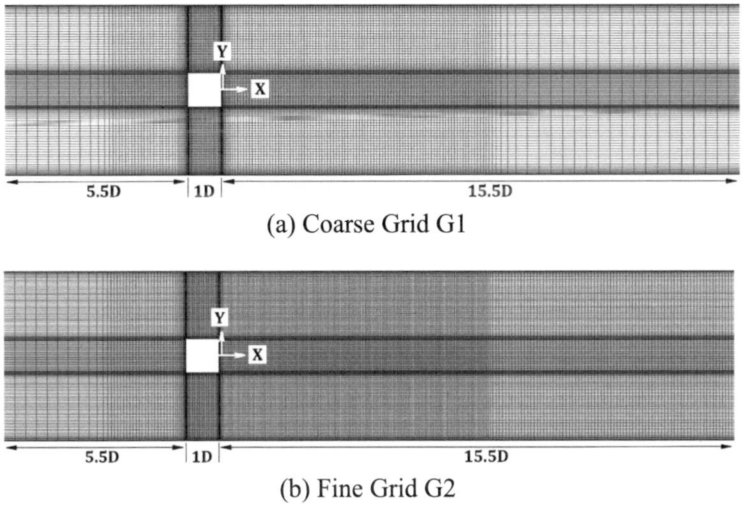

Fig. 2.28 Grid distribution in the X–Y plane of the computational domain for the square cylinder

fluctuations predicted by both hybrid methods on the fine grid are closely aligned with the experimental results.

The mean streamlines of the flow field are shown in Fig. 2.30. Downstream of the square cylinder, a pair of counter-rotating vortices forms, and recirculation zones are observed near the upper and lower surfaces of the cylinder. The flow separated from the sharp leading edges of the square cylinder does not reattach to the upper and lower surfaces but instead interacts with the downstream separation vortices to create a large separation zone, which contrasts with the separation patterns seen around circular cylinders. The VLES method's prediction of the separation zone size closely matches the LES results, while the DDES method significantly underestimates the length of the separation zone. Furthermore, refining the grid has a minimal effect on the predicted size of the separation zone for both the VLES and DDES methods.

The mean velocity, turbulent fluctuations, and Reynolds stresses in the flow field are sampled at $x/h = 1,\ 3.5,\ 6.0$ and 8.5. Their distributions are portrayed in figures from Fig. 2.31, 2.32, 2.33, 2.34 and 2.35. Given that the mean flow field is symmetric in the Y direction, only the distribution of variables in the lower half of the channel is presented.

In Figs. 2.31 and 2.32, the streamwise and normal mean velocities predicted by the VLES method closely match the experimental data and LES results. The DDES method generally performs well but shows deviations near the centerline at $x/h = 1$. Figure 2.31 illustrates that the momentum loss near the leeward face of the square cylinder quickly recovers downstream. Figure 2.32 demonstrates that the mean normal velocity near the square cylinder is significantly influenced by the pair of counter-rotating vortices behind it, with their impact diminishing rapidly further downstream. The VLES method precisely estimate the velocity recovery

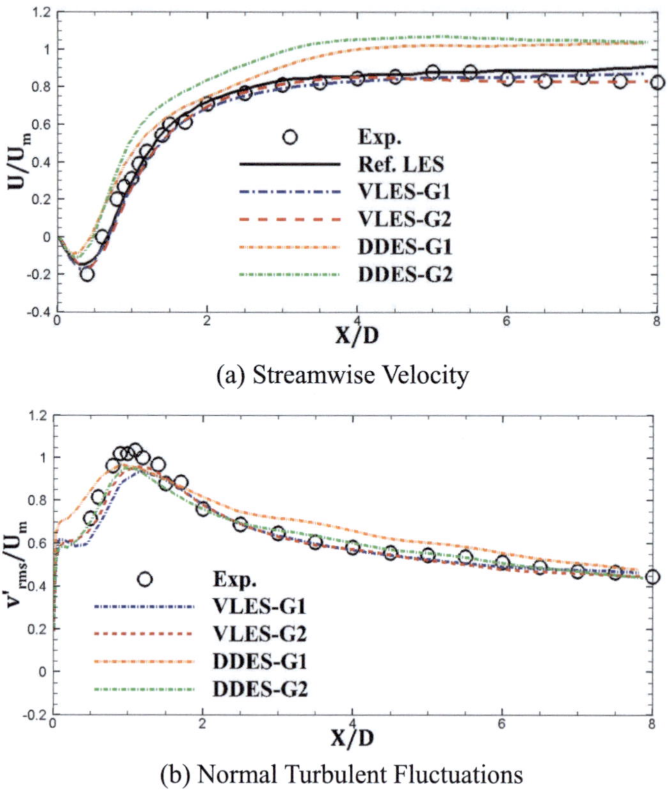

Fig. 2.29 Distribution of mean streamwise velocity and normal turbulent fluctuations along the centerline downstream of the square cylinder

in the wake region, where small parameter changes can significantly affect flow development. In Fig. 2.33, there are some discrepancies between the numerically predicted streamwise velocity fluctuations and the experimental results. Nevertheless, in Fig. 2.34, the predicted normal velocity fluctuations closely align with the experimental measurements. Near the square cylinder, the maximum amplitude of normal velocity fluctuations exceeds that of streamwise velocity fluctuations, which indicates strong vertical momentum transport caused by the separated vortices from the upper and lower surfaces of the square cylinder. Figure 2.35 shows the Reynolds stress distributions. The VLES method's predictions align very well with the experimental data at all streamwise positions, while the DDES method shows deviations at $x/h = 3.5$. The figure also reveals that the Reynolds stress is primarily influenced by the normal mean velocity near the square cylinder.

Table 2.6 presents the characteristics of the square cylinder wake as determined by different methods, including the root mean square of the lift coefficient C_{Lrms}, the mean drag coefficient $C_{D\mathrm{mean}}$, the root mean square of the drag coefficient C_{Drms}, the Strouhal number St, and the recirculation region length L_r/D. The Strouhal number

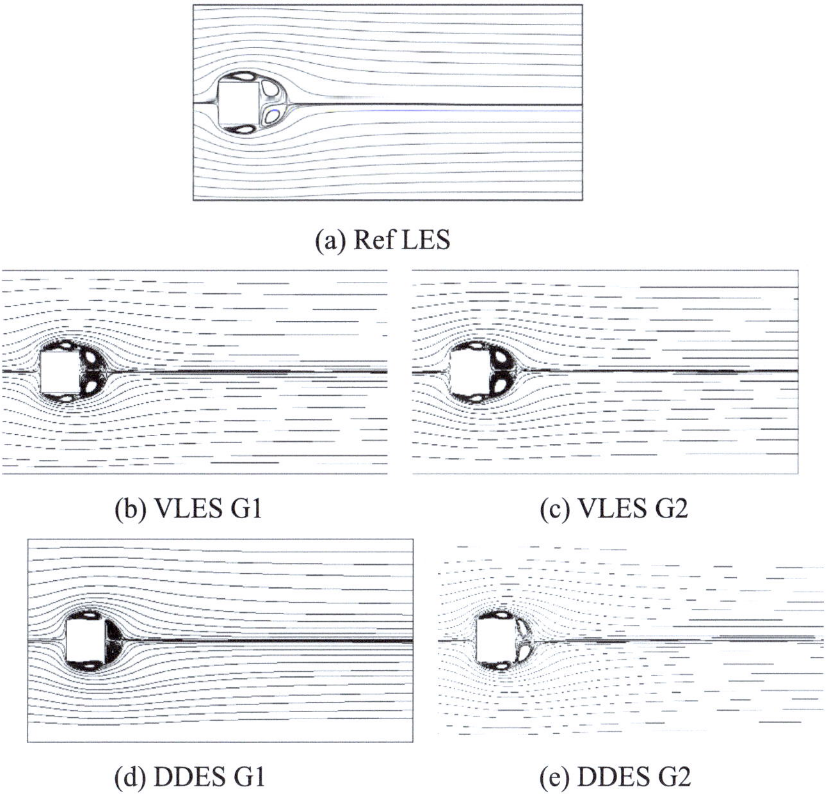

Fig. 2.30 Streamline distribution around the square cylinder

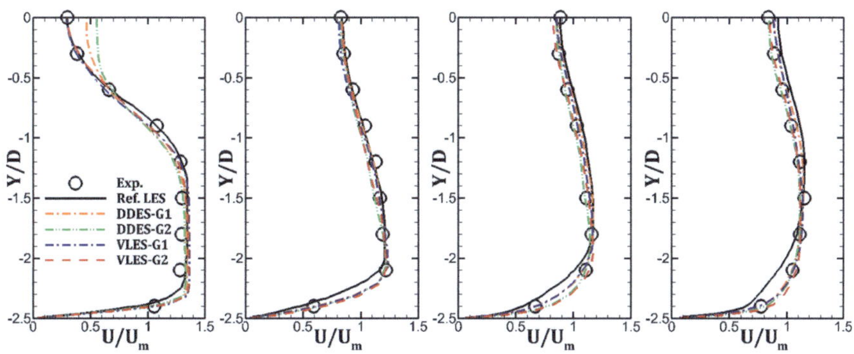

Fig. 2.31 Streamwise mean velocity profiles at $x/h = 1,\ 3.5,\ 6.0,$ and 8.5 positions

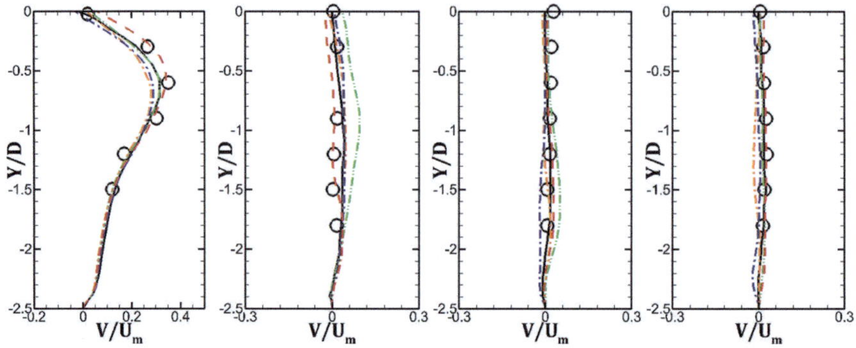

Fig. 2.32 Normal mean velocity profiles at $x/h = 1,\ 3.5,\ 6.0,$ and 8.5 positions

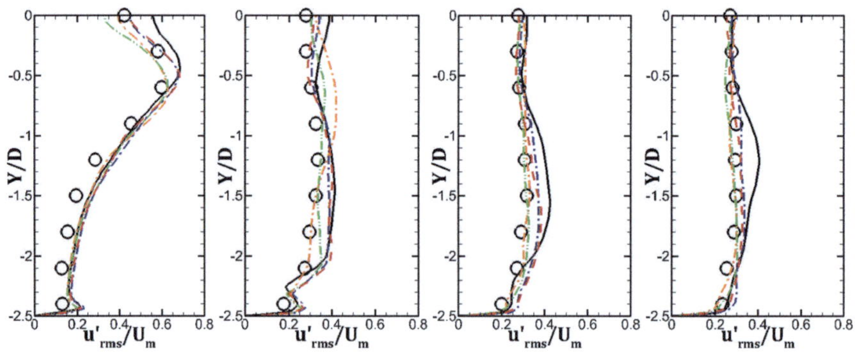

Fig. 2.33 Root mean square distribution of streamwise velocity fluctuations at $x/h = 1,\ 3.5,\ 6.0,$ and 8.5 positions

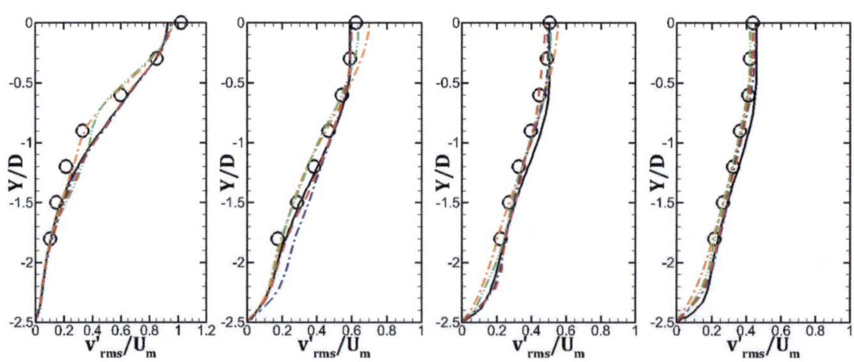

Fig. 2.34 Root mean square distribution of normal velocity fluctuations at $x/h = 1,\ 3.5,\ 6.0,$ and 8.5 positions

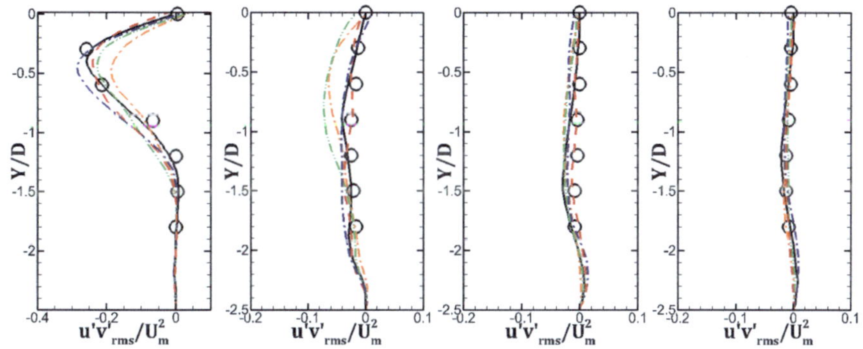

Fig. 2.35 Root mean square distribution of reynolds stresses at $x/h = 1, 3.5, 6.0$, and 8.5 positions

Table 2.6 Comparison of square cylinder wake characteristics

	Exp	Ref. LES	VLES-G1	VLES-G2	DDES-G1	DDES-G2
C_{Lrms}	–	2.06	2.079	2.053	2.739	2.434
$C_{D\text{mean}}$	–	2.76	2.929	2.825	2.852	2.709
C_{Drms}	–	0.49	0.403	0.436	0.927	0.674
St	0.13	0.124	0.144	0.138	0.161	0.153
L_r/D	0.6	0.67	0.71	0.69	0.46	0.49

is derived from the lift frequency, the side length of the square cylinder, and the mean velocity, i.e., $St = fD/U_m$. Predictions from both the VLES and DDES methods are higher than experimental values and LES data. Grid refinement leads to decreased predictions from both hybrid methods. The VLES method predicts a slightly larger recirculation region length compared to experimental values and LES data, while the DDES method predicts a significantly smaller recirculation region length. For the root mean square of the lift and drag coefficients, the VLES method results are closer to the LES data, whereas the DDES method results are significantly higher. Using a coarse grid, both hybrid methods predict a $C_{D\text{mean}}$ larger than the LES data; in contrast, with grid refinement, the $C_{D\text{mean}}$ values predicted by the hybrid methods decrease.

Figure 2.36 represents the iso-surface distribution of the flow field around the square cylinder predicted by the VLES method, with the flow field colored according to the local streamwise velocity. The image shows that the separation vortices detach from both the upper and lower surfaces of the square cylinder. As these vortices move downstream, they are constrained by the channel walls, which causes the formation of separation bubbles along the walls. Comparing results from two different grid resolutions, the refined grid enables the VLES method to capture more detailed turbulent vortex structures, while the overall developmental trend of the wake remains consistent.

Fig. 2.36 Distribution of iso-surfaces of q-criterion for square cylinder flow around different grids predicted by the VLES method

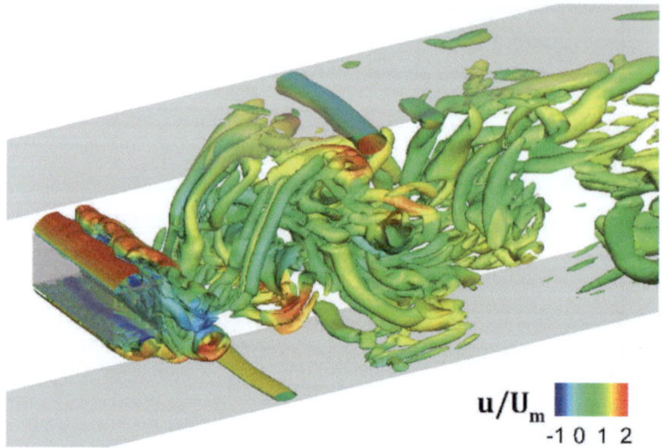

(a) Coarse Grid G1

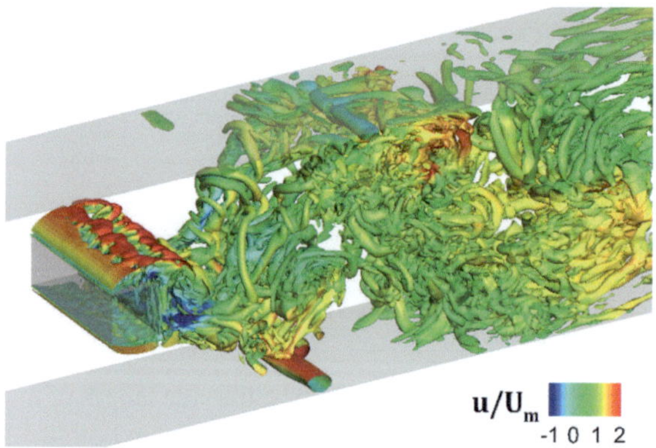

(b) Fine Grid G2

In this square cylinder flow case, the SST-VLES method achieves higher simulation accuracy than the SST-DDES method, attaining a level comparable to that of the LES method.

2.5 Summary

This chapter outlines the structural features of the Very Large Eddy Simulation (VLES) method and suggests improvements to address the shortcomings of the Speziale-VLES method. The SST-VLES method was developed based on the $k-\omega$ SST model. To mitigate the excessive decay of Reynolds stress in the VLES method and improve performance in flows with a wide range of length scales, the turbulent integral scale was used to replace the Kolmogorov length scale. Model parameters were re-derived through dimensional analysis and equilibrium turbulence theory to ensure the proper activation of the LES mode. To address the mode's gray area issue, the filtering scale was redefined, and a boundary layer protection function was introduced, which can enhance the SST-VLES method's accuracy in simulating boundary layers and separated flows. In various test cases, the SST-VLES method's predictions align well with experimental data or DNS and LES results. The Taylor–Green vortex simulation results demonstrate that the SST-VLES method can accurately capture turbulent structures and their dissipation processes. Comparative results from cases such as the 30P30N multi-element airfoil, single cylinder, and square cylinder flows indicate that the SST-VLES method exhibits excellent capability in simulating turbulent separation flows.

Chapter 3
Prediction Model for Bypass Transition and Separation-Induced Transition

Abstract This chapter presents a novel indicator capable of characterizing pressure gradients and sensing turbulent fluctuation intensity at the boundary layer edge, while satisfying Galilean invariance. Leveraging direct numerical simulation (DNS) data, prediction models for both bypass transition and separation-induced transition are developed in RANS framework. The accuracy and universality of these models have been rigorously validated through extensive wind tunnel experiments on benchmark test cases.

3.1 Introduction

Under the framework of the hybrid RANS/LES method, the prediction of transition locations in the boundary layer using the VLES method is based on the underlying RANS transition model. To advance a LES method capable of accurately capturing the transition process, it is imperative first to develop an efficient, robust, and scalable RANS transition prediction model suitable for extensive parallel computing.

Laminar flow on wings or turbomachinery blades can greatly reduce friction drag, aerodynamic heating, and aerodynamic noise. As a result, transition prediction is of paramount importance in the design of aircraft and turbomachinery. Blades in turbomachinery typically operate under conditions of high turbulence intensity. Therefore, the study of transition mechanisms in such environments is essential. In high turbulence conditions, the length and amplitude of the strong Klebanoff disturbances increase progressively downstream, ultimately resulting in breakdown through the formation of turbulent spots. This process, governed by streak instability [81, 82], bypasses the conventional instability mechanisms of natural transition and is thus referred to as bypass transition [85]. Furthermore, the frequent occurrence of adverse pressure gradients in turbomachinery can induce laminar boundary layer separation, thereby necessitating the precise prediction of laminar separation bubbles and their subsequent reattachment.

Researchers have amassed extensive transition data and flow mechanism analyses through wind tunnel experiments and DNS data. Based on these findings, the

© The Author(s), under exclusive license to Springer Nature Singapore Pte Ltd. 2025
J. Xu et al., *CFD-Compatible RANS/LES Modeling of Transitional and Separated Flows*, https://doi.org/10.1007/978-981-96-6886-1_3

objective is to develop a prediction model within the RANS framework that can precisely describe bypass transition and capture its essential dynamics. Motivated by the advancements in single-equation transition models, this study aims to formulate a novel, efficient, and streamlined single-equation transition model specifically designed for high-turbulence intensity environments, with a focus on turbomachinery flows.

This chapter first introduces a novel local indicator to depict the effects of turbulence intensity and pressure gradient. In the aftermath, a new critical Reynolds number for transition is proposed, which is capable of predicting bypass transition and also provides a reasonable simulation of transition processes on flat plates with low turbulence intensity. Additionally, to address the transition induced by laminar separation bubbles, the model is adjusted to account for the influence of pressure gradient and turbulence intensity. This enables the precise simulation of the flow reattachment process.

3.2 Development of the Transition Model

3.2.1 *Transport Equation for the Intermittency Factor*

This study integrates the intermittency factor transport equation with Menter's $k-\omega$ SST model [130] to simulate boundary layer transition process. The intermittency factor γ is calculated using the transport equation [79, 80, 160–163], which is formulated as follows:

$$\frac{\partial(\rho\gamma)}{\partial t} + \frac{\partial(\rho u_j\gamma)}{\partial x_j} = P_\gamma - E_\gamma + \frac{\partial}{\partial x_j}\left[\sigma_\gamma(\mu + \sigma_f\mu_t)\frac{\partial\gamma}{\partial x_j}\right], \tag{3.1}$$

where the source term is given by

$$P_\gamma = \rho c_{a1} S F_{\text{onset}}(1 - c_{e1}\gamma)[-\ln(1-\gamma)]^{\frac{1}{3}}, \tag{3.2}$$

$$E_\gamma = \rho c_{a2}\Omega F_{\text{turb}}\,\gamma(c_{e2}\gamma - 1). \tag{3.3}$$

In the equation, ρ represents density, S denotes the shear strain rate, and Ω indicates vorticity. The values of the model coefficients are listed in Table 3.1. Furthermore, F_{onset} is used to activate the production term, while F_{turb} regulates the destruction/ re-laminarization source term.

Building on previous research [68, 79, 80, 160, 161, 163] and calibration of test cases, the following transition onset detection function has been established.

Table 3.1 Coefficients for the Intermittency Factor Transport Equation Model

σ_γ	σ_f	c_{a1}	c_{a2}	c_{e1}	c_{e2}
0.2	25.0	80.0	0.06	1.0	50.0

$$F_{\text{onset1}} = 2.0 \cdot \tanh\left(\left(\frac{\text{Re}_V}{2.193\text{Re}_{\text{crit}}}\right)^{30.0}\right), \tag{3.4}$$

$$F_{\text{onset2}} = \min\left(\max\left(F_{\text{onset1}}, F_{\text{onset1}}^4\right), 2.0\right), \tag{3.5}$$

$$F_{\text{onset3}} = \max\left(1 - \left(\frac{R_T}{2.0}\right)^3, 0\right), \tag{3.6}$$

$$R_T = \frac{\rho k}{\mu\omega}, \tag{3.7}$$

$$F_{\text{onset}} = \max(F_{\text{onset2}} - F_{\text{onset3}}, 0), \tag{3.8}$$

$$F_{\text{turb}} = \mathrm{e}^{-(R_T)^4}, \tag{3.9}$$

Here, R_{T} is the viscosity ratio, $\text{Re}_V = \frac{\rho S d^2}{\mu}$ is the vorticity Reynolds number, and Re_{crit} represents the critical Reynolds number at which flow instability begins.

The triggering function in the above equation is based on the model developed by Langtry and Menter. The form of this intermittency factor equation has been extensively validated and is highly reliable. In high turbulence intensity environments, the establishment of an accurate and effective transition criterion is of paramount importance.

3.2.2 Indicator Factors and Transition Criteria

The key to modeling bypass transition lies in providing an appropriate critical Reynolds number under varying turbulence intensities and pressure gradients. Extensive parameter analysis has revealed that the product of the viscosity ratio (the ratio of eddy viscosity to molecular viscosity) and the pressure gradient parameter can simultaneously describe the flow's turbulent fluctuation intensity, pressure gradient, and dissipation characteristics. For this reason, the indicator factor is defined as follows:

$$T_\omega = R_T H_c, \tag{3.10}$$

From the above equation, the following dimensional analysis can be obtained:

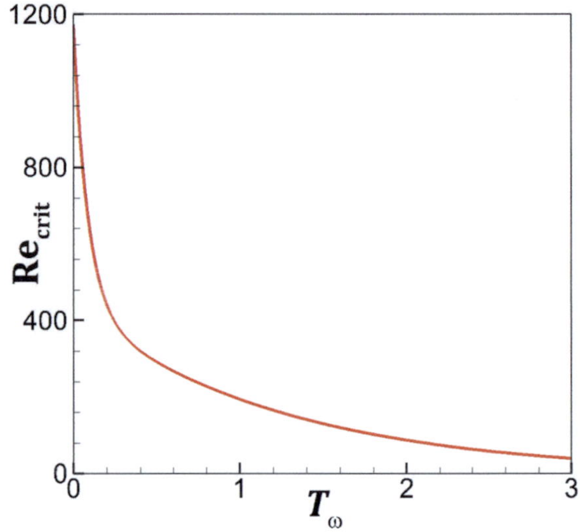

Fig. 3.1 Relationship between the critical transition reynolds number and the indicator factor

$$R_T H_c = \frac{\nu_t}{\nu}\frac{|S|d}{u} \sim \frac{1}{\nu}\frac{\nu_t \frac{\partial u}{\partial d}}{\omega} \sim \frac{\left|\overline{u'v'}\right|}{\nu\omega} \tag{3.11}$$

In this equation, the variables ν and ν_t represent the laminar kinematic viscosity and turbulent kinematic viscosity, respectively. The local dimensionless pressure gradient parameter H_c, introduced by Coder [164], effectively describes the pressure gradient within the boundary layer.

Drawing on the work of Durbin and Ge, this study introduces a new critical transition Reynolds number, $\mathrm{Re}_{\mathrm{crit}}$, for bypass transition in high-turbulence environments. The form of $\mathrm{Re}_{\mathrm{crit}}$ is provided in Eq. 3.12. Figure 3.1 demonstrates the variation of the critical transition Reynolds number with the indicator factor T_ω. As turbulence intensity and pressure gradient increase, T_ω rises, the critical transition Reynolds number decreases, resulting in an earlier onset of transition.

$$\mathrm{Re}_{\mathrm{crit}} = \max\left(\min\left(743.1e^{-11.61T_\omega} + 430.3e^{-0.7982T_\omega}, 1200\right), 40\right). \tag{3.12}$$

3.2.3 *Correction for Transition Induced by Laminar Separation Bubbles*

Wings and turbine blades frequently encounter extensive regions characterized by adverse pressure gradients, which predispose the laminar boundary layer to the formation of laminar separation bubbles. To add to this, the free-stream turbulence intensity dramatically influences the transition process driven by these separation bubbles [4, 165–170]. As the turbulence intensity in the incoming flow increases, the size of the laminar separation bubble diminishes, and the point of separation onset shifts downstream. In turn, free-stream turbulence intensity must be a critical consideration in the development of transition models. To exactly predict the transition of separation bubbles and simulate the subsequent reattachment process, it is essential to refine the intermittency factor. The effective intermittency factor γ_{eff} is defined as follows:

$$\gamma_{\mathrm{eff}} = \max(\gamma, \gamma_{\mathrm{sep}}) \tag{3.13}$$

Here, γ_{sep} denotes the separation correction intermittency factor. For transitions induced by separation bubbles, the model is adjusted according to the methodology of Langtry and Menter [67, 68]. The key approach involves allowing the local intermittency factor to exceed 1.0 when the laminar boundary layer separates. This adjustment facilitates a substantial generation of turbulent kinetic energy k, which in turn promotes the earlier reattachment of the separated flow [68]. To avoid excessive turbulent kinetic energy production, the separation bubble correction intermittency factor γ_{sep} is capped at an upper limit of 2.0.

Analysis of the similarity solutions for the laminar boundary layer indicates that a separation bubble will form when the adverse pressure gradient is less than or equal to $\beta = -0.1988$, which corresponds to a shape factor $H_{12} = 4.0$. Here, β is the dimensionless Hartree pressure gradient parameter [171]. The relationship between the shape factor H_{12} and the pressure gradient parameter β is depicted in Fig. 3.2a. To determine the occurrence of the laminar separation bubble, a shape factor switching function T_b is defined. For a critical shape factor H_{12} of 4.0, if H_{12} is less than 4.0, the flow remains attached to the wall, and T_b is set to 0.0. Conversely, if H_{12} exceeds 4.0, flow separation occurs, and T_b is set to 1.0.

$$T_b = \begin{cases} 1.0 & H_{12} \geq 4.0, \\ 0.0 & H_{12} < 4.0. \end{cases} \tag{3.14}$$

In order to address the challenge of determining the non-local shape factor H_{12} using local variables, Menter et al. [69] initially proposed a local shape factor H_L. This methodology was subsequently refined by Coder [172].

$$H_L = \frac{d^2}{\mu}\left[\nabla(\rho\vec{u} \cdot \nabla d) \cdot \nabla d\right], \tag{3.15}$$

In this context, the symbol ∇d denotes the unit vector normal to the wall. The formulation of the local shape factor bears a close resemblance to the Hartree and Thwaites shape factors. Through the analysis of similarity solutions for the laminar boundary layer, it is evident that the integral shape factor H_{12} and the local shape factor H_L exhibit a monotonic functional relationship, as depicted in Fig. 3.2b.

$$H_{12} = \min(\max(0.0935H_L + 2.382, 2.17), 8.0). \tag{3.16}$$

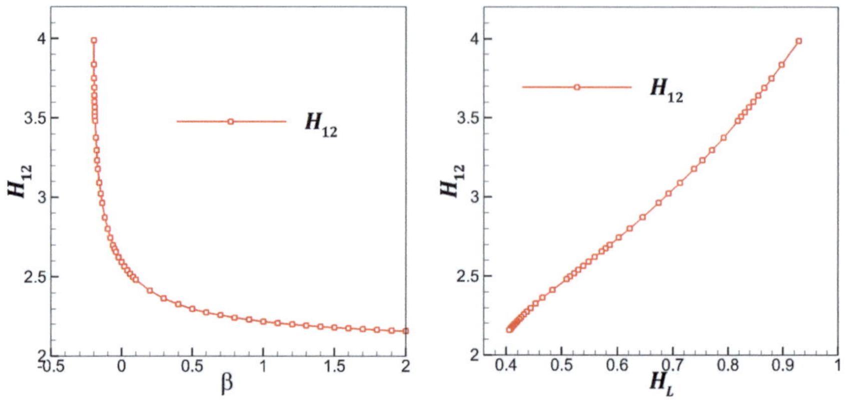

(a) the Relationship Between H_{12} and β (b) the Relationship Between H_{12} and H_L

Fig. 3.2 Curves of shape factor H_{12} as a function of β and H_L

In the above context, the functional relationship is derived through parameter analysis and function fitting of the similarity solutions for the laminar boundary layer. Furthermore, compared to the relationship proposed by Coder [172] in 2017, this function yields values of the shape factor H_{12} that align more closely with theoretical predictions. Equation 3.16 ensures that the shape factor for a zero-pressure gradient flat plate is 2.59. The specified upper limit of 8.0 and lower limit of 2.17 ensure that the shape factor values remain within a physically plausible range. In regions experiencing adverse pressure gradients, a more precise estimation of the momentum thickness Reynolds number is necessary. Therefore, the derived local shape factor H_{12} is utilized to compute the momentum thickness Reynolds number. The relationship between the vorticity Reynolds number $\mathrm{Re}_\theta = \frac{\rho_e \theta U_e}{\mu_e}$ and the momentum thickness Reynolds number is expressed by Eqs 3.16 and 3.17 [79].

$$\mathrm{Re}_\theta = \frac{\mathrm{Re}_V}{2.193F(H_{12})}. \tag{3.17}$$

$$F(H_{12}) = \min(2.2,\max(0.8475H_{12} - 1.195,0.5)) \tag{3.18}$$

In this equation, the function $F(H_{12})$ serves to constrain the value of Re_θ in the separation region, thereby ensuring that it does not exceed a reasonable range.

For the simulation of transition in attached flows, Eppler [173] proposed the $H_{32} - R_{\delta_2}$ relationship, which demonstrated robust prediction capabilities. Subsequently, Dini [174] extended this methodology to the simulation of transition in separated boundary layers, thereby developing the momentum thickness Reynolds number $\mathrm{Re}_{\theta s}$, which is applicable to laminar separation bubbles, as depicted in Fig. 3.3.

$$\mathrm{Re}_{\theta s} = \exp\left[125(H_{32} - 1.573)^2 + 18.4H_{32} - 21.74\right] \tag{3.19}$$

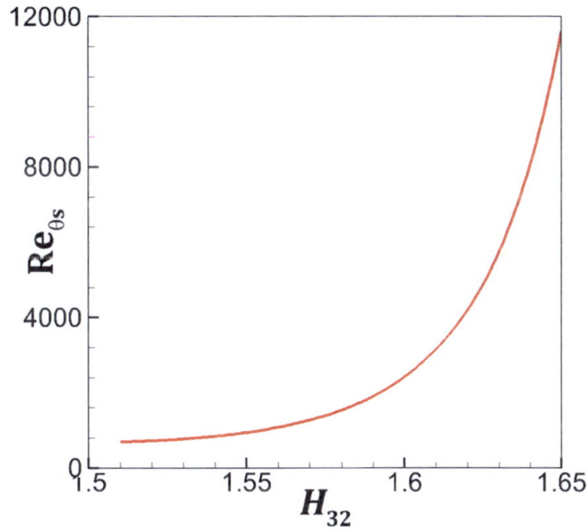

Fig. 3.3 Relationship between momentum thickness reynolds number and shape factor H_{32}

In this equation, the shape factor H_{32} represents the ratio of the boundary layer's energy loss thickness to its momentum loss thickness. Figure 3.4 illustrates how the shape factor H_{32} varies with the pressure gradient parameter β and the local shape parameter H_L. Similar to H_{12}, H_{32} can be calculated using H_L by analyzing the similarity solution for the laminar boundary layer and applying function fitting techniques.

$$H_{32} = \max\big(\min\big(0.1816e^{-0.2928H_L} + 1.47e^{0.002351H_L}, 1.62\big), 1.515\big) \tag{3.20}$$

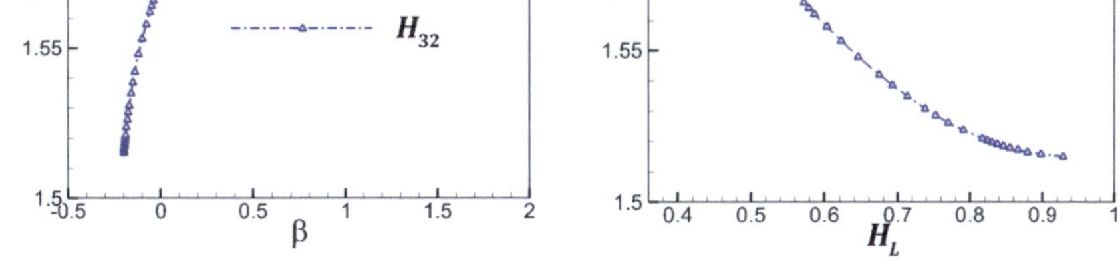

(a) the Relationship Between H_{32} and β (b) the Relationship Between H_{32} and H

Fig. 3.4 Variation of shape factor H_{32} with β and H_L

Turbulence intensity and pressure gradient have a significant impact on the formation, development, and dissipation of laminar separation bubbles, thereby affecting the transition process induced by the separation bubble. To account for the effects of turbulence intensity and pressure gradient, the empirical function f_{modi} is applied to modify $\text{Re}_{\theta s}$, with model parameters calibrated using both experimental and DNS data.

Accordingly, the intermittency factor for predicting the transition of laminar separation bubbles is updated as follows:

$$\gamma_{\text{sep}} = \min(T_b B_L B_{\text{re}}, 2.0) \tag{3.21}$$

$$B_L = \max\left(\text{Re}_\theta - \frac{\text{Re}_{\theta s}}{f_{\text{modi}}}, 0\right) \tag{3.22}$$

$$f_{\text{modi}} = \max(\min(f_{\text{modi1}}, f_{\text{modi2}}), 1.0), \tag{3.23}$$

$$f_{\text{modi1}} = \min\left(1.959\text{e}^{93.86 T_\omega} - 1.05\text{e}^{-1959 T_\omega}, 8.3\right) \tag{3.24}$$

$$f_{\text{modi2}} = \min\left(4.39\text{e}^{0.4147 T_\omega} - 1.739\text{e}^{-29.29 T_\omega}, 8.3\right), \tag{3.25}$$

$$B_{\text{re}} = \text{e}^{-\left(\frac{R_T}{10}\right)^3} \tag{3.26}$$

When the viscosity ratio reaches a sufficient level to induce flow reattachment, B_{re} sets γ_{sep} to zero, thereby negating the correction. In high-turbulence environments, the boundary layer (B_L) can regulate the size of the separation bubble. As it reaches a considerable value in the vicinity of the laminar separation bubble, γ_{sep} experiences a rapid increase and exceeds 1.0, resulting in the generation of substantial turbulent kinetic energy and ultimately leading to flow reattachment. Figure 3.5 illustrates the relationship between the modified empirical function f_{modi} and the indicator T_ω. As turbulence intensity and pressure gradient increases, T_ω rises and the value of f_{modi} also increases, resulting in an earlier transition onset.

3.2.4 Coupling Between Transition and Turbulence Models

Finally, to simulate the physical effects of boundary layer transition, the effective intermittency factor γ_{eff} derived from the above equations is coupled into the turbulent kinetic energy source term of Menter's $k - \omega$ SST turbulence model.

$$\frac{\partial(\rho k)}{\partial t} + \frac{\partial(\rho u_j k)}{\partial x_j} = \tilde{P}_k - \tilde{D}_k + \frac{\partial}{\partial x_j}\left[(\mu + \sigma_k \mu_t)\frac{\partial k}{\partial x_j}\right], \tag{3.27}$$

In the above equation, the values of P_k and D_k are constrained by γ_{eff}.

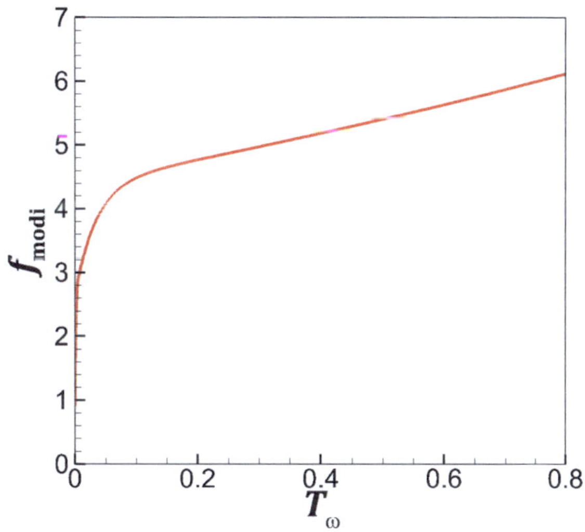

Fig. 3.5 The relationship between the function f_{modi} and the indicator factor T_ω

$$\tilde{P}_k = \gamma_{eff} P_k, \tag{3.28}$$

$$\tilde{D}_k = \min(\max(\gamma_{eff}, 0.1), 1.0) D_k, \tag{3.29}$$

The transport equation for the specific dissipation rate is consistent with the original SST turbulence model. Furthermore, to ensure the activation of the $k - \omega$ model in the laminar boundary layer and transition region, the blending function F_1 is modified as follows:

$$R_y = \frac{\rho y \sqrt{k}}{\mu}, \tag{3.30}$$

$$F_3 = e^{-\left(\frac{R_y}{120}\right)^8}, \tag{3.31}$$

$$F_1 = \max\left(F_{1,\,orig}, F_3\right). \tag{3.32}$$

In the equation, $F_{1,\,orig}$ denotes the blending function from the original SST turbulence model.

At this stage, a single-equation transition model based on local variables has been fully developed. The reliability of the proposed transition model will be demonstrated through several classic test cases.

3.3 Case Validation

3.3.1 *Flat Plate Without Separation*

In this section, the new model is evaluated using the Schubauer and Klebanoff (S&K) flat plate experiments [175] and the T3 series flat plate experiments [176, 177] conducted by the European Research Community on Flow, Turbulence, and Combustion (ERCOFTAC). These experiments are widely recognized as benchmark cases for the validation of transition models. The inlet boundary conditions for these cases are specified in Table 3.2. The T3A, T3B, and T3A-flat plate cases represent pressure-gradient-free flat plates with free-stream turbulence intensities exceeding 1%, thus primarily exhibiting bypass transition. In contrast, the S&K case, also a pressure gradient-free flat plate, has a lower intensity of free flow turbulence, resulting in a natural transition. This configuration facilitates the assessment of the model's ability to predict transition behavior in low-turbulence environments. The T3C flat plate case comprises a series of plates subjected to varying streamwise pressure gradients, achieved by modifying the upper surface contour of the wind tunnel. This introduces changes in boundary layer development. As the wind tunnel's inlet velocity decreases and the Reynolds number reduces, the transition onset on the plate shifts from a favorable pressure-gradient region near the leading edge to an adverse pressure-gradient region near the trailing edge. Thus, the T3C flat plate serves as a classic test case for examining the effects of pressure gradients and boundary layer history on transition prediction.

For a new transition model, it is essential to validate grid convergence. To this end, grid convergence tests were conducted on the T3A flat plate using three distinct mesh configurations, as summarized in Table 3.3. The dimensionless height of the first mesh layer, y^+, was maintained below 1.0 for all three meshes. Figure 3.6 presents a comparison of the surface friction coefficients obtained with these meshes. The "Coarse" mesh resulted in a transition location that was upstream compared to the "Medium" and "Fine" meshes. The transition locations predicted by the "Medium"

Table 3.2 Inlet boundary conditions for the T3 series and S&K flat plates

Case	U_{inlet} (m/s)	Tu_{inlet} (%)	R_T	$\nu(\text{m}^2/\text{s}^{-1})$
T3A	5.4	3.3	12.0	1.5×10^{-5}
T3B	9.2	6.2	100.0	1.5×10^{-5}
T3A-	19.8	0.874	8.7	1.5×10^{-5}
S&K	50.1	0.3	1.0	1.5×10^{-5}
T3C1	6.4	8.0	50.0	1.5×10^{-5}
T3C2	5.5	2.8	10.0	1.5×10^{-5}
T3C3	4.1	3.0	6.0	1.5×10^{-5}
T3C5	9.2	3.5	17.0	1.5×10^{-5}

Table. 3.3 Mesh information used for grid convergence analysis

	Coarse	Medium	Fine
Number of grid points in the normal direction	61	97	141
Number of grid points in the streamwise direction	121	181	289
Normal grid growth rate	1.15	1.10	1.05

and "Fine" meshes were almost identical and in good agreement with the experimental data, indicating that further mesh refinement does not significantly alter the transition location. This confirms the grid convergence of the transition model. Consequently, the "Fine" mesh was selected for the simulations of the flat plate cases. Figure 3.7 illustrates the mesh configurations used for the T3 series flat plate cases. The computational domain features a slip wall at the upper boundary, while the lower boundary is a slip wall from $x = -0.2$ m to $x = 0.0$ m and a no-slip wall for $x \geq 0$. Mesh refinement is applied at the plate's inlet.

The sensitivity of the transition model to turbulence intensity and viscosity ratio was also validated using the T3A flat plate. The simulation results shown in Fig. 3.8 exhibit significant shifts in the transition location with changes in turbulence intensity and viscosity ratio, following a logical trend. Specifically, as the turbulence intensity increases, the transition location moves upstream; as the viscosity ratio decreases, the transition location shifts downstream. This clearly demonstrates that in high-turbulence intensity environments, the proposed indicator and transition criteria can sensitively and accurately capture the influence of turbulence intensity on the transition location.

The inlet boundary conditions were set according to Table 3.2, and the distributions of free-stream turbulence intensity and velocity along the streamwise direction

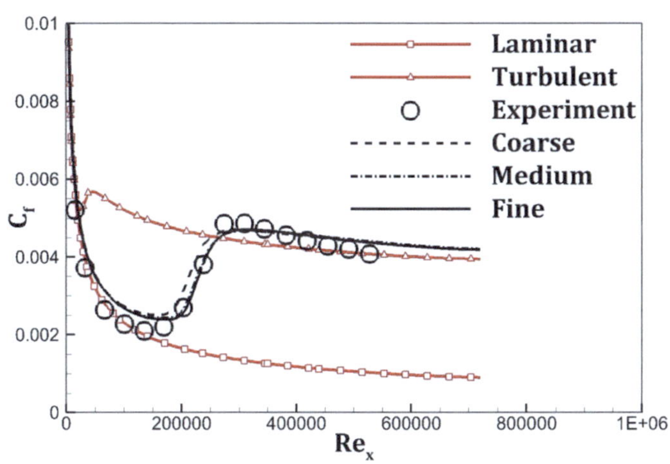

Fig. 3.6 Validation of grid convergence for the T3A flat plate

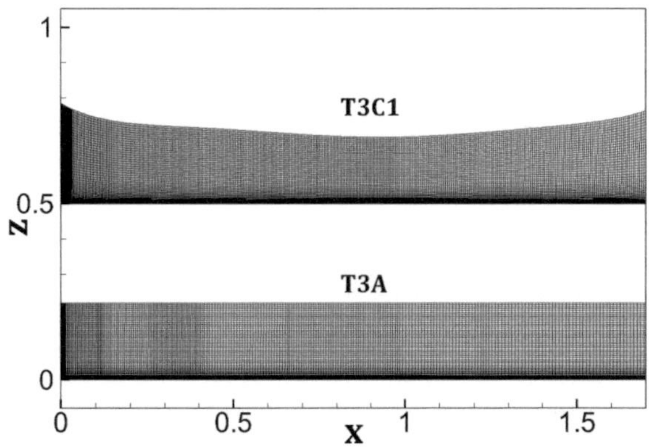

Fig. 3.7 Mesh distribution for the T3 series of flat plates

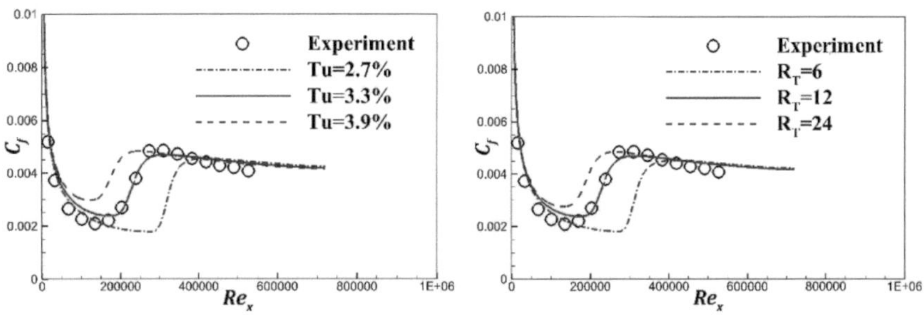

(a) Effect of Turbulence Intensity (b) Effect of Viscosity Ratio

Fig. 3.8 Variation in transition location on the T3A flat plate under different incoming flow conditions

for the flat plate flow fields were calculated. Figures 3.9 and 3.10 illustrate the variations in free-stream turbulence intensity with streamwise position for each flat plate case, while Fig. 3.11 shows the changes in free-stream velocity along the streamwise position for the T3C series flat plates. The simulation results from the transition model are in strong agreement with the experimental data, indicating that the flow conditions in the numerical simulations accurately reproduce the experimental conditions. Based on these boundary conditions, the distribution of the surface friction coefficient on the flat plates was determined, as shown in Figs. 3.12 and 3.13. Overall, the transition model's predictions closely match the experimental measurements, which demonstrates that the developed transition model effectively captures the onset and progression of transition in flows with varying pressure gradients and turbulence intensities. The model also provides sufficiently accurate predictions of surface friction.

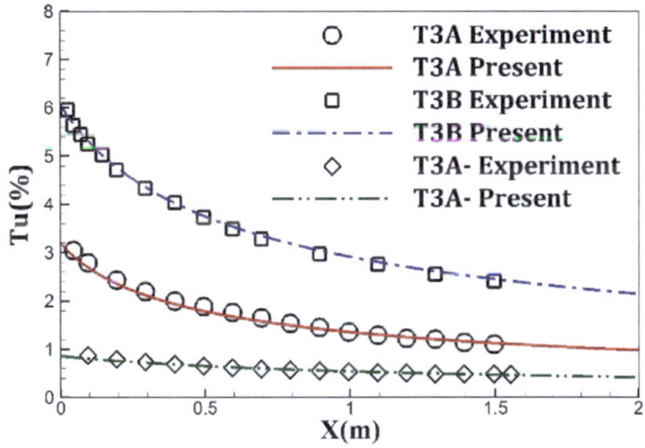

Fig. 3.9 Distribution of free-stream turbulence intensity for the T3A, T3B, and T3A- flat plate cases

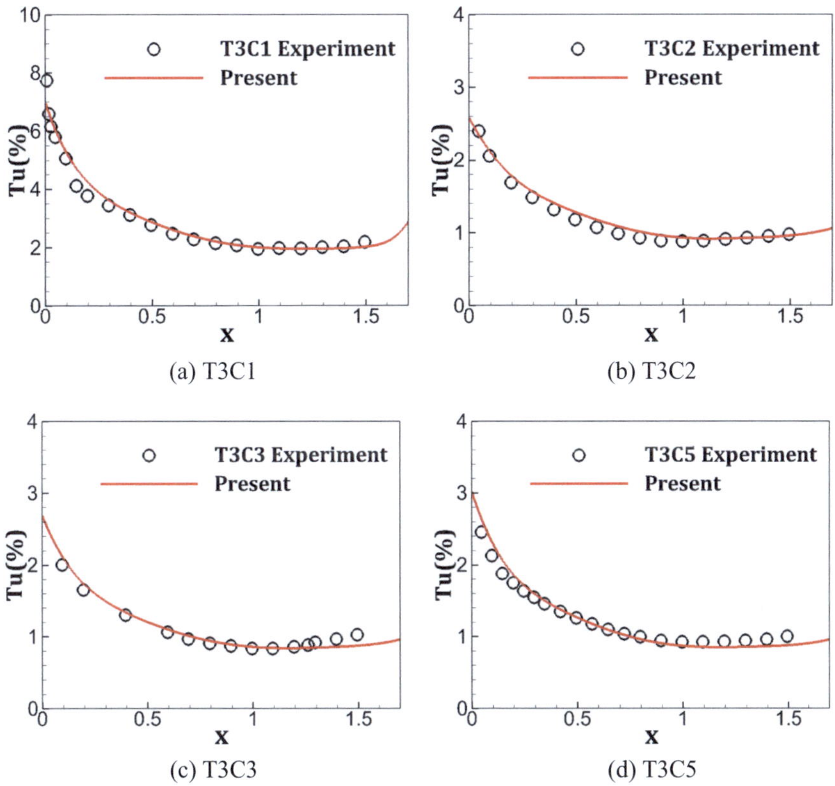

Fig. 3.10 Variation of free-stream turbulence intensity along the streamwise position for the T3C series flat plates

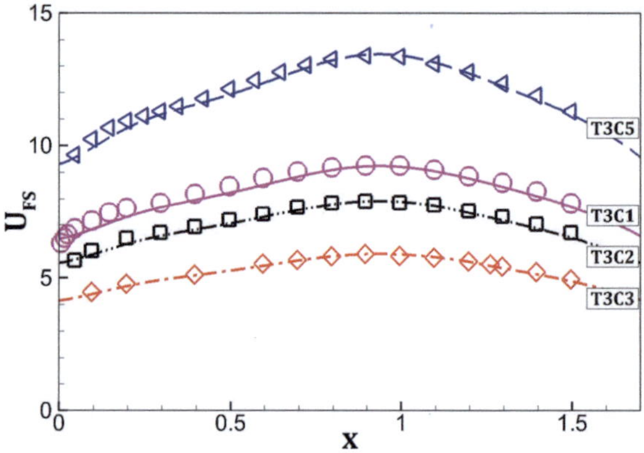

Fig. 3.11 Variation of free-stream velocity along the streamwise position for the T3C series flat plates

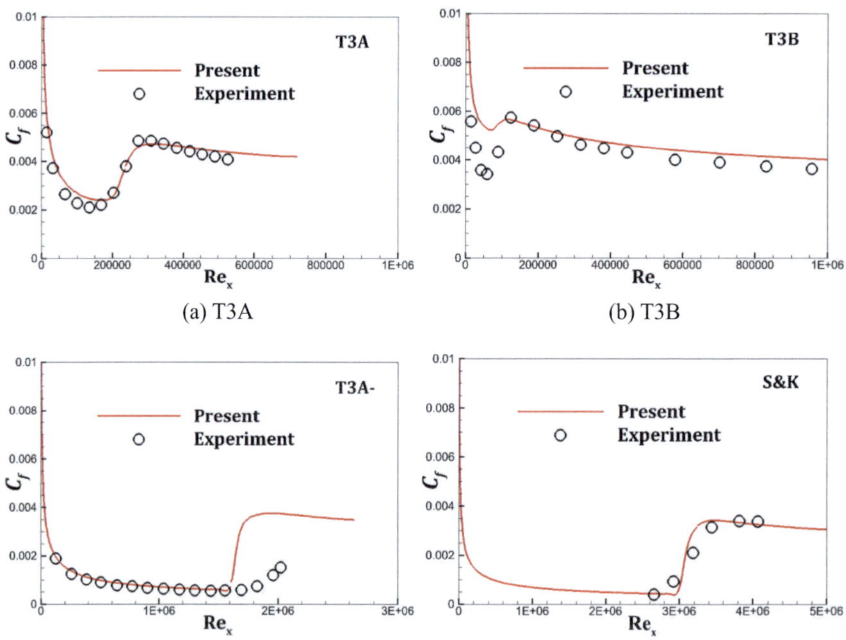

(a) T3A

(b) T3B

(c) T3A-

(d) S&K

Fig. 3.12 Distribution of surface friction coefficient for the T3 series flat plates without pressure gradient

3.3.2 Flat Plate with Pressure Gradient and Separation

This section aims to validate the new transition model's accuracy in predicting transition induced by laminar separation bubbles by simulating flows over multiple flat plates with pressure gradients. In addition, it examines the influence of free-stream

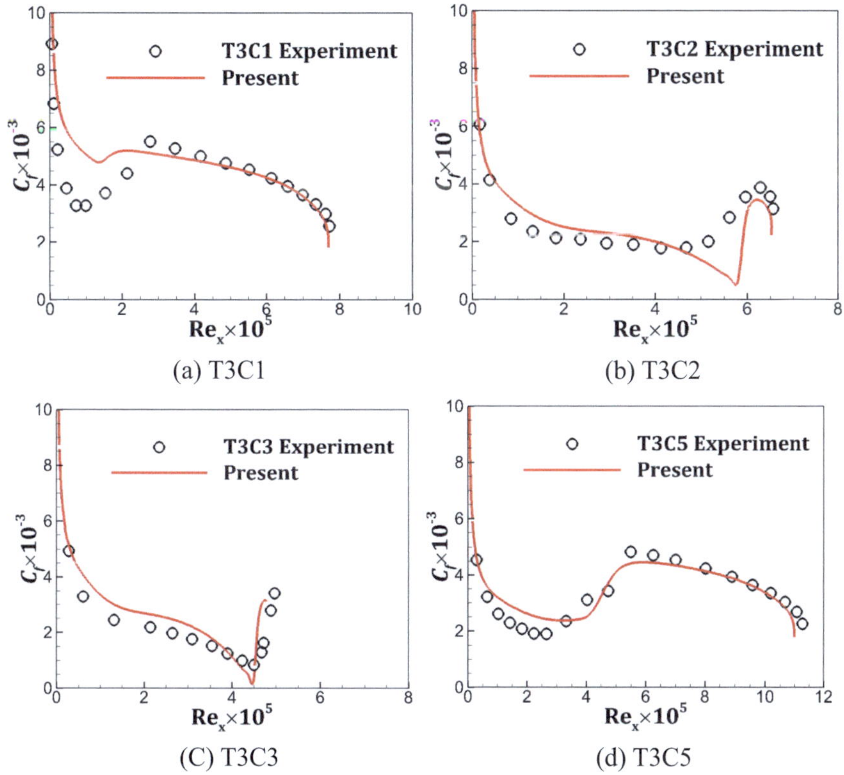

Fig. 3.13 Distribution of surface friction coefficient for the T3C series flat plates

turbulence intensity. The first flat plate case is T3C4, part of the T3 series [176, 177]. Following this, numerical simulations are conducted for the flat plate flow experiments by Lou and Hourmouziadi [178] (L&H Flat). Three different inlet boundary conditions for free-stream turbulence intensity are selected, with comparisons made to the numerical results by Lardeau et al. [46] and Ge et al. [84]. Furthermore, the flat plate flow experiments by Hultgren and Volino [179] (H&V Flat) are utilized to further validate the transition model's accuracy. Table 3.4 presents the inlet boundary conditions for these cases, where the Reynolds number (Re) is calculated based on the plate length (L) and the incoming flow velocity (U_{in}). The boundary conditions and mesh distribution for the flat plate computational domain are illustrated in Fig. 3.14.

In the T3C4 flat plate test case, the boundary conditions were set according to the provided table, resulting in calculated incoming flow turbulence intensity and velocity distribution that closely matched the experimental data, as shown in Fig. 3.15a. This scenario features a low incoming flow velocity and a low Reynolds number. Due to the adverse pressure gradient, the laminar boundary layer separates near the outlet. Following a separation-induced transition process, the flow field reattaches as a turbulent boundary layer. Although the separation bubble's location

Table 3.4 Inlet boundary conditions for flat plates with pressure gradients and separation

Case	U_{in} (m/s)	Tu_{in} (%)	R_T	$\nu(m^2/s^{-1})$	Re
T3C4	1.3	2.9	3.5	1.5×10^{-5}	173,400
L&H Flat A	3.4	2.4	3.0	1.5×10^{-5}	60,000
L &H Flat B	3.4	4.0	7.0	1.5×10^{-5}	60,000
L&H Flat C	3.4	5.8	10.0	1.5×10^{-5}	60,000
H&V Flat A	11.0	0.2	1.0	1.5×10^{-5}	50,000
H&V Flat B	11.0	7.0	10.0	1.5×10^{-5}	50,000
H&V Flat C	11.0	0.2	1.0	1.5×10^{-5}	300,000
H&V Flat D	11.0	7.0	10.0	1.5×10^{-5}	300,000

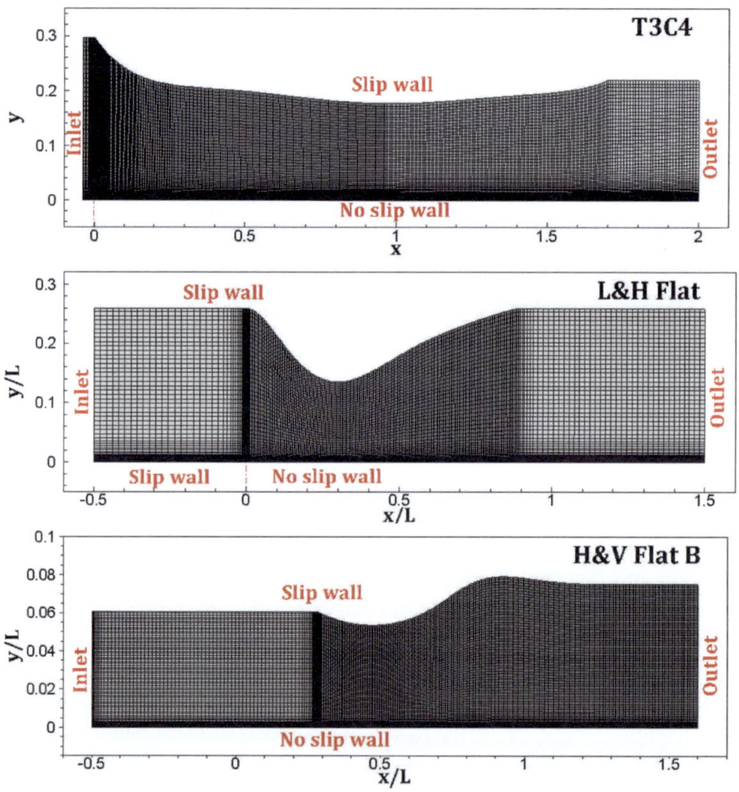

Fig. 3.14 The grid arrangement and boundary condition settings for a flat plate with an adverse pressure gradient

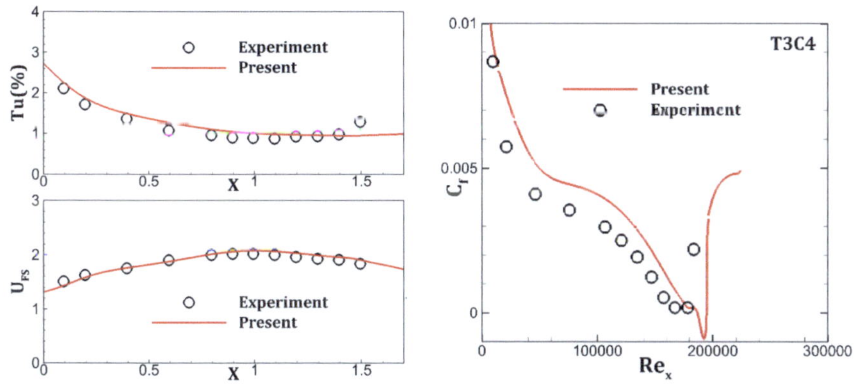

(a) Incoming Flow Turbulence Intensity and Velocity Distribution

(b) Surface Friction Coefficient

Fig. 3.15 Results of the T3C4 flat plate test case

predicted by the numerical simulation is slightly downstream, the overall distribution of the predicted surface friction coefficient aligns well with the experimental results (Fig. 3.15b). Figure 3.16 illustrates the turbulence intensity, intermittency factor, and turbulent kinetic energy around the laminar separation bubble. Above the separation bubble, there is a region where the effective intermittency factor is greater than 1 but less than 2. The model enhances the production term of turbulent kinetic energy to promote the reattachment of the separated flow. Thus, increased turbulent kinetic energy can be observed in the flow field downstream of the separation bubble. This indicates that the new transition model accurately captures the separation-induced transition process.

Lardeau et al. [46] employed the LES method to investigate the L&H flat plate and obtained incoming flow turbulence intensities of 1.0%, 1.5%, and 2.0%. The current transition model was applied to simulate these three conditions, with boundary conditions detailed in Table 3.4. To align the streamwise decay of turbulence intensity with the results from Lardeau et al., the inlet turbulence intensities for the RANS transition model in this study were respectively set at 2.4%, 4.0%, and 5.8%, as depicted in Fig. 3.17a. This method of specifying inlet conditions differs from that of the LES method but is analogous to the RANS boundary condition settings utilized by Ge et al. [84]. By ensuring that the turbulence intensity decay distribution is consistent, the overall flow field state can be accurately maintained. Figure 3.17b and d present the distribution of surface friction along the flat plate.

The figures indicate that in the L&H Flat B and C test cases, the predicted locations of the separation and reattachment points align well with the results from Lardeau et al. and Ge et al. The surface friction coefficient distributions obtained from the two RANS transition models are very similar, though both exhibit slight discrepancies compared to the LES data. In the L&H Flat A test case, where the incoming turbulence intensity is relatively low, the reattachment point predicted by the transition model appears slightly earlier than that predicted by the LES method.

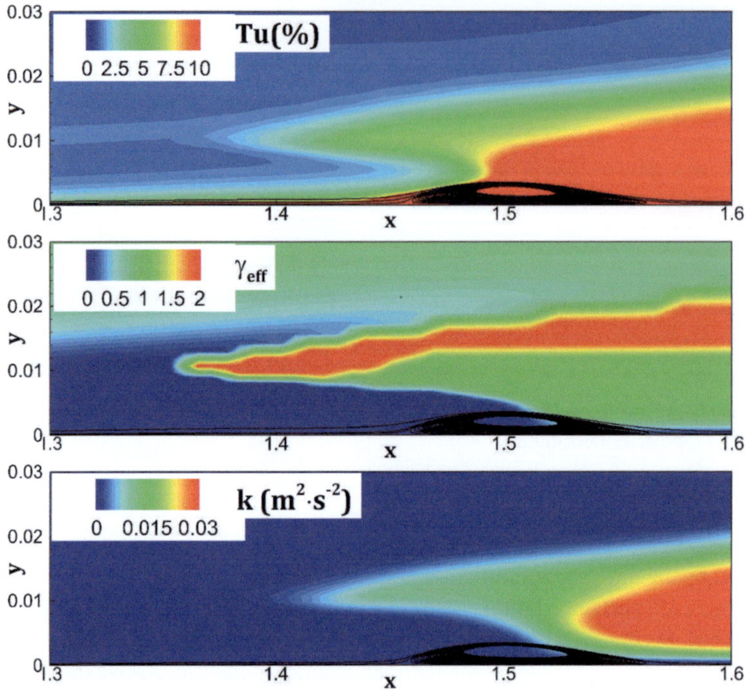

Fig. 3.16 Contour plots of turbulence intensity, intermittency factor, and turbulent kinetic energy near the separation bubble in the T3C4 flat plate test case

Hultgren and Volino [179] conducted experimental studies on flat plate flow under varying Reynolds numbers and incoming turbulence intensities. Subsequently, Suzen et al. [66] and Menter et al. [69] used RANS transition models to perform numerical simulations on these flat plate test cases. In this subsection, we conduct a comparative analysis of two test cases, each with different Reynolds numbers (50,000 and 300,000) and incident turbulence intensities (0.2% and 7.0%). According to Menter et al., the experimental inlet turbulence intensities of 0.2% and 7.0% correspond to test section turbulence intensities of 0.2% and 2.5% in the test section respectively. Therefore, these test section turbulence intensity values were adopted as boundary conditions for the numerical simulations, with other inlet conditions outlined in Table 3.4. All upper boundary shapes are derived from the study of Suzen et al. [66]. To create a proper laminar boundary layer before the flow enters the pressure region, the leading edge of the flat plate was positioned at $x/L = 0.0$, and the computational domain boundary was extended to $x/L = -0.5$. Figure 3.14 shows the grid distribution and boundary settings for the H&V flat plate.

Figure 3.18 represents the effects of Reynolds number and incoming turbulence intensity on the separation bubble. As the Reynolds number or the intensity of the incoming turbulence increases, the size of the separation bubble decreases and may even disappear, as shown in Fig. 3.18d. The surface friction coefficient distributions

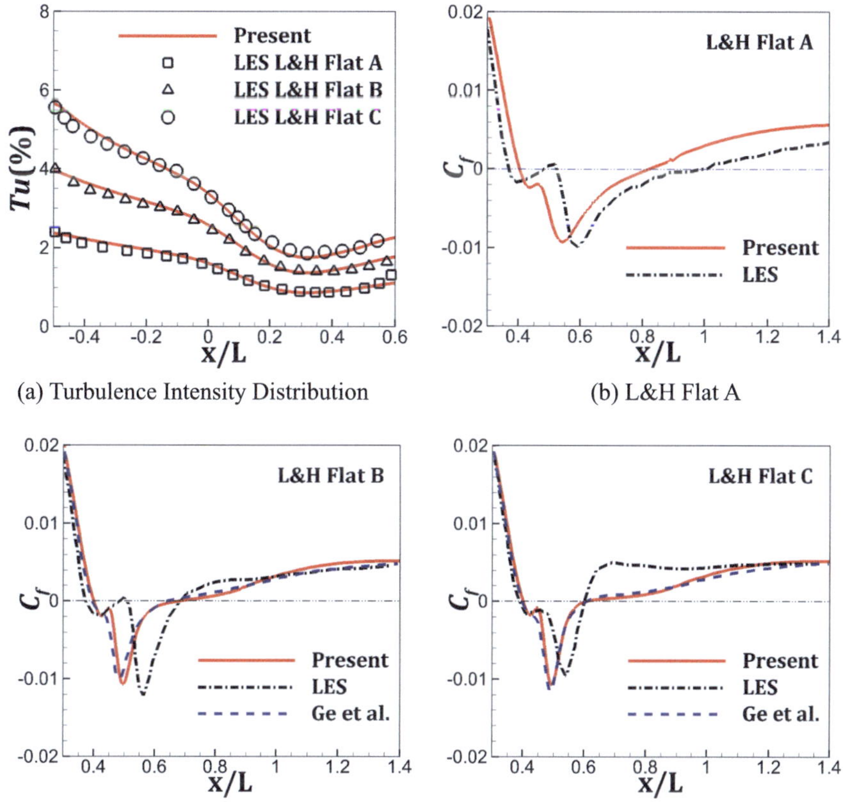

(a) Turbulence Intensity Distribution

(b) L&H Flat A

(c) L&H Flat B

(d) L&H Flat C

Fig. 3.17 Streamwise decay of turbulence intensity and surface friction coefficient distribution for the L&H flat plate

for the flat plate under various conditions are presented in Fig. 3.19. Results from the γ-model [69] and the $\gamma - \overline{Re_{\theta t}}$ model [68] are also plotted for comparison with the current model's outcomes. In the H&V Flat A test case, the separation bubble size predicted by the current model is slightly smaller than that predicted by the other two transition models and the experimental data. In the H&V Flat D test case, all three numerical simulation methods observed the disappearance of the separation bubble. In this case, the transition location predicted by the current model is slightly delayed compared to the other two models and is more in line with the experimental results. In addition, in the H&V Flat B and C test cases, the transition point predicted by the current model occurs slightly earlier, and the reattachment point occurs earlier as well, more closely matching the experimental results compared to the other models.

Overall, the model can effectively capture the influence of Reynolds number and incoming turbulence intensity on bubble separation and transition. As the influencing parameters are changed, the predicted size of the laminar separation

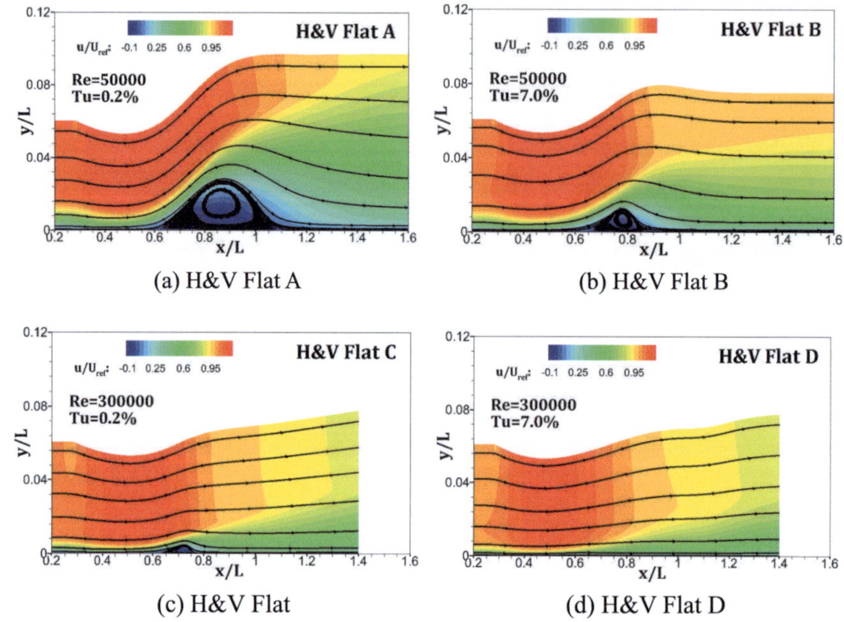

(a) H&V Flat A (b) H&V Flat B

(c) H&V Flat (d) H&V Flat D

Fig. 3.18 Streamline plots of the flow field for the H&V flat plate

bubble and the transition location undergo logical and consistent variations, which demonstrates the model's robustness in simulating the effects of these parameters.

3.3.3 T3L Series Separation Flat Plate

The T3L test cases are a component of the flat plate experimental studies on transitional flow conducted by the ERCOFTAC Special Interest Group (SIG) [176, 180]. In this part, numerical simulations are carried out for four specific conditions: T3L1, T3L2, T3L3, and T3L4. The experiments were conducted on a flat plate with a semi-circular leading edge, with a diameter of $D = 10$ mm. The free stream angle of attack and the pressure gradient were both set to zero, while the free stream velocity was maintained at $U_{\text{in}} = 5$ m/s. The incident turbulence intensity varied among different groups, with the specific inlet conditions outlined in Table 3.5. The Reynolds number is computed by the leading-edge diameter and the free stream velocity, and the turbulence intensity is measured at $x = 0.006$, they are also presented in the table.

The computational domain for this test case is defined as follows: $x_{\min} = -0.3$ m, $x_{\max} = 0.3$ m, $y_{\min} = 0$ m, $y_{\max} = 0.218$ m. Figure 3.20 depicts the grid distribution for the entire flow field and the leading-edge region of the flat plate. The mesh comprises approximately 47,000 cells, meanwhile the dimensionless height of

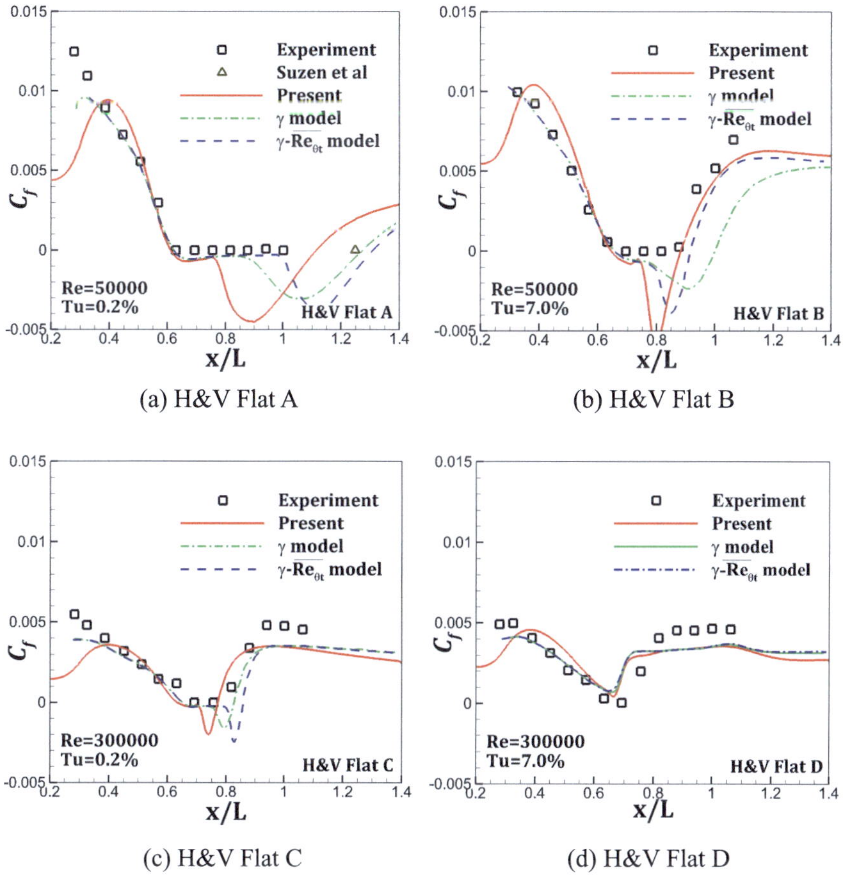

(a) H&V Flat A

(b) H&V Flat B

(c) H&V Flat C

(d) H&V Flat D

Fig. 3.19 Surface friction coefficient distribution for the H&V flat plate

Table 3.5 Inlet boundary conditions for T3L test cases

	Case	T3L1	T3L2	T3L3	T3L4
	Re_D	3293	3293	3293	3293
	U_{inlet}	5.0	5.0	5.0	5.0
	Tu_{LE} (%)	0.17	0.63	2.39	5.34

the first grid layer y^+ is less than 0.2. Slip wall boundary conditions are applied to the upper boundary of the flow field and the lower boundary preceding the leading edge of the flat plate (Fig. 3.20).

Figure 3.21 shows the distributions of the effective intermittency factor and turbulent kinetic energy in the flow field under various inlet conditions. It is evident that laminar separation begins after the flow has crossed the leading-edge curve, with

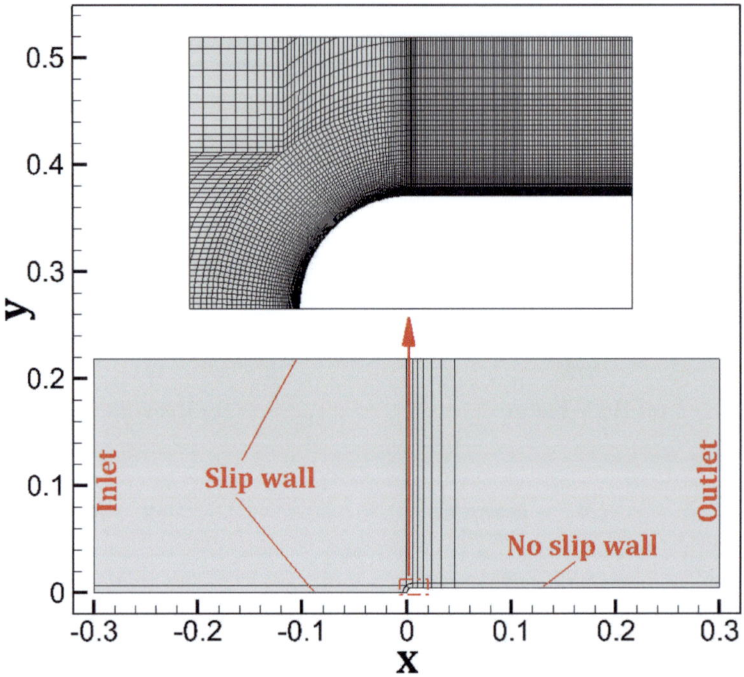

Fig. 3.20 Computational domain and grid distribution of the leading-edge region for the T3L flat plate

the separation onset location remaining consistent across different inlet conditions. As the intensity of the free stream turbulence increases, the transition point moves upstream, causing the peak turbulent kinetic energy to occur earlier. Concurrently, the transition point shifts from 37.74 to 21.51 mm. This result demonstrates the effectiveness of incorporating the indicator factor T_{ω} to refine the transition criteria and the intermittency factor transport equation. This adjustment allows the model to precisely reflect the impact of turbulence intensity on transition induced by laminar separation bubbles. With an increase in turbulence intensity, the γ_{sep} value within the separation bubble rises rapidly, which causes earlier flow reattachment and a smaller separation bubble.

The reattachment locations predicted by the transition model are presented in Table 3.6 and compared with experimental data, as well as the computational results from Vicedo et al. [181] and Wang et al. [76]. In low turbulence intensity conditions, the model's predicted reattachment points demonstrate a high degree of agreement with the experimental observations. Nonetheless, in high turbulence intensity environments, the predicted size of the separation bubble is found to be slightly larger than that measured experimentally. The surface friction coefficient distribution along the flat plate is shown in Fig. 3.22. A negative C_f value indicates the presence of reverse flow and flow separation. In the experiments, C_f was set to 0 within the separation zone. As depicted in the accompanying figure, the predictions exhibit a

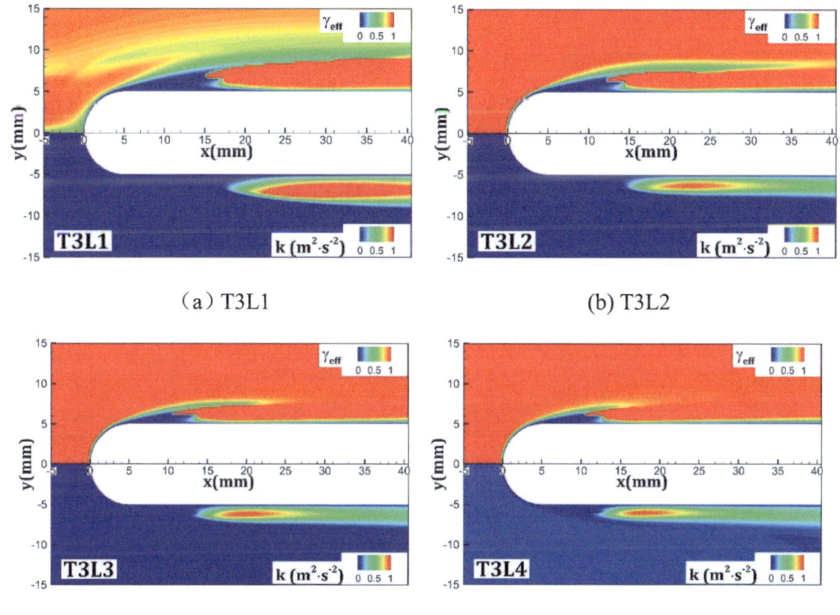

(a) T3L1

(b) T3L2

(c) T3L3

(d) T3L4

Fig. 3.21 Contour plots of effective intermittency factor and turbulent kinetic energy near the leading edge for T3L flat plate test cases

high degree of alignment with the corresponding experimental data. In the T3L4 test case, the computed friction coefficient in the turbulent region is found to be slightly higher than the experimental values.

To compare the flow details within the separation bubble, Fig. 3.23 presents the velocity profile distributions at different streamwise locations for incoming turbulence intensities of $Tu = 0.63$ and 2.39%. The vertical axis represents the distance normal to the wall.

Figure 3.23 shows that as the laminar separation bubble develops, the backflow velocity within the bubble first increases and then decreases. After reattachment, the flow transitions to a turbulent boundary layer velocity profile. In test case T3L3, the simulated separation bubble size is slightly larger than the experimental measurements. Figure 3.24 reveals the variation in turbulent kinetic energy distribution along

Table 3.6 Reattachment locations for T3L test cases

Case	T3L1	T3L2	T3L3	T3L4
Experimental x_r(mm)	40	27	23	21
Predicted x_r(mm)	37.74	28.4	24.53	21.51
Vicedo et al. [181] x_r(mm)	–	28.3	24.3	21.2
Wang et al. [76] x_r(mm)	39.0	29.2	24.2	20.5

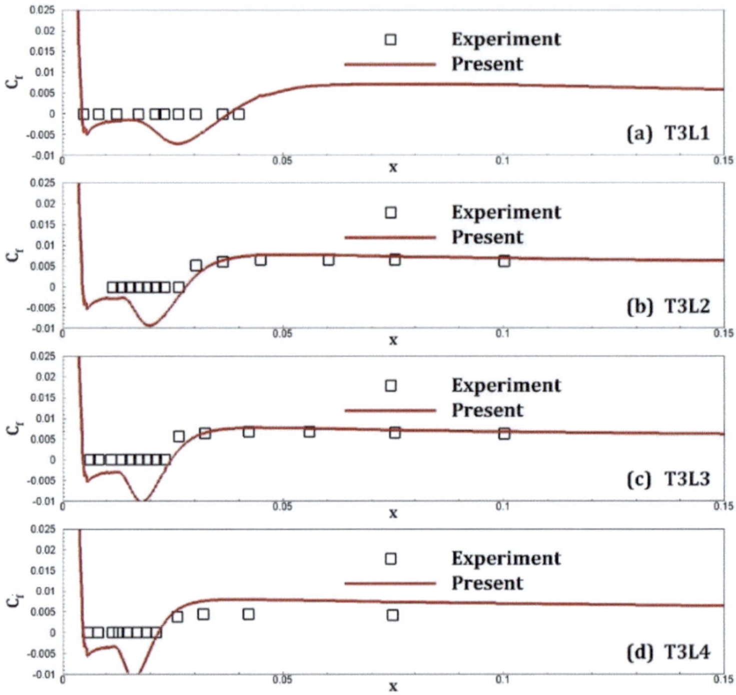

Fig. 3.22 Comparison of surface friction coefficient distributions for T3L test cases

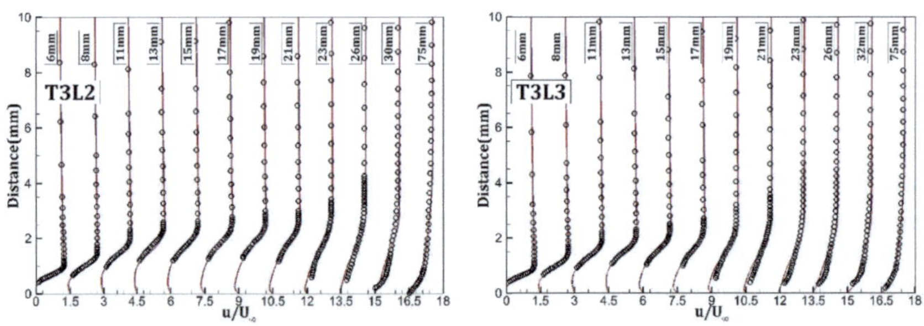

(a) T3L2 (b) T3L3

Fig. 3.23 Velocity profile distribution at different streamwise locations for the T3L flat plate case

the flat plate at different streamwise locations for different incident turbulence intensities. As represented in Fig. 3.24b, the predicted transition point is approximately at $x = 15$ mm, marked by the onset of increased turbulent kinetic energy. In the pre-transition region, turbulent kinetic energy remains minimal. After transition, the predicted turbulent kinetic energy distribution closely matches the experimental data. However, the predicted transition point is somewhat downstream of the experimental

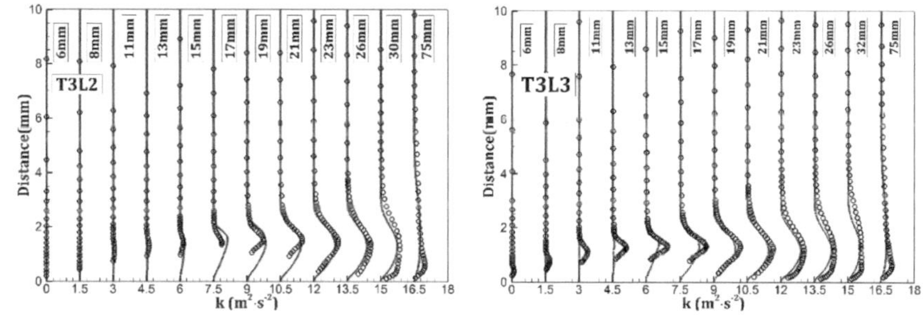

(a) T3L2 (b) T3L3

Fig. 3.24 Turbulent kinetic energy distribution at different streamwise locations for the T3L case

results, resulting in a delayed increase in turbulent kinetic energy that also affects the reattachment process. Figure 3.24a shows that the predicted location of the increased turbulent kinetic energy is consistent with the experimental observations. Overall, the transition model demonstrates robust accuracy in characterizing the separation bubble size, transition location, and reattachment behavior under varying turbulence intensity conditions.

3.3.4 Pak-B Low-Pressure Turbine Blade

The loads experienced by the Pratt and Whitney Pak-B low-pressure turbine blade are analogous to those encountered by modern aircraft engine rotors [182]. Transition occurs within the free shear layer of the separation bubble on the blade's suction side. The location at which transition commences in the free shear layer is a determining factor in whether the separation bubble can reattach as a turbulent boundary layer, which in turn affects the potential for blade stall. This particular transition process presents a considerable challenge for numerical simulations. This section will demonstrate how the new transition model can be used to address the aforementioned simulation challenge.

Huang et al. [183] conducted a comprehensive experimental investigation of the Pak-B turbine cascade, they examined a range of inlet conditions with varying Reynolds numbers and turbulence intensities. The geometric schematic is presented in Fig. 3.25. The Reynolds numbers were calculated based on the axial chord length of the blades and the freestream velocity, with values primarily set at 50,000, 75,000, and 100,000.

The free-stream flow angle was maintained at 35°. The inlet turbulence intensities include 0.08%, 2.35%, and 6.0% (corresponding to turbulence intensities at the blade leading edge of 0.08%, 1.6%, and 2.85%, respectively). Figure 3.26 presents the setup of boundary conditions within the computational domain and the mesh distribution around the blade leading edge.

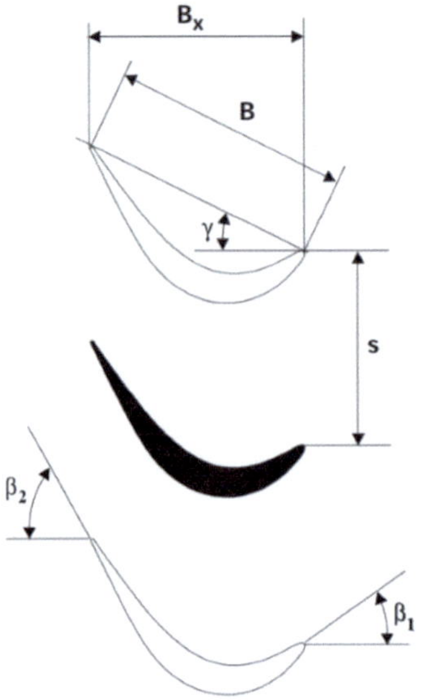

Axial Chord (B_x)	75mm
True Chord (B)	83.4mm
Blade Spacing (S)	66.8m
Stagger Angle (γ)	26°
Inlet Angle (β_1)	35°
Outlet Angle (β_2)	60°

Fig. 3.25 Schematic of the Pak-B cascade experimental geometry

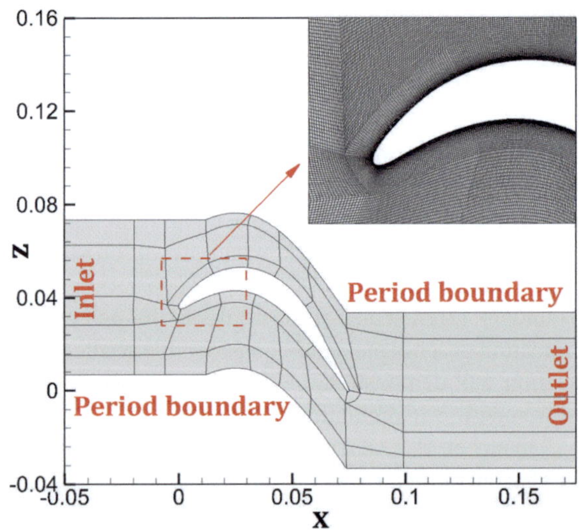

Fig. 3.26 Computational domain of the Pak-B blade and mesh distribution near the leading edge

Figure 3.27 depicts the blade loading under disparate free-stream turbulence intensities and Reynolds numbers. As either the Reynolds number or free-stream turbulence intensity increases, the separation bubble size decreases and the length of the pressure plateau shortens. The computational results demonstrate a high degree of correlation with the experimental data across all four conditions, which substantiates the assertion that the new model is an effective means of capturing the effects of variations in Reynolds number and free-stream turbulence intensity. This indicates the model's capability to accurately reflect the impact of these variables on the size of the laminar separation bubble and the subsequent turbulent reattachment process.

As illustrated in Fig. 3.28, laminar separation bubbles are present on both the suction and pressure sides of the blade. The transition occurring within the free shear layer of these separation bubbles leads to an increase in turbulent kinetic energy, which subsequently promotes flow reattachment. Figure 3.29 provides contour plots of turbulent kinetic energy and turbulence intensity for the Pak-B low-pressure cascade. Downstream of the transition induced by the separation bubbles, there is a rapid increase in both turbulent kinetic energy and turbulence intensity, which is attributed to the growth of the turbulent boundary layer.

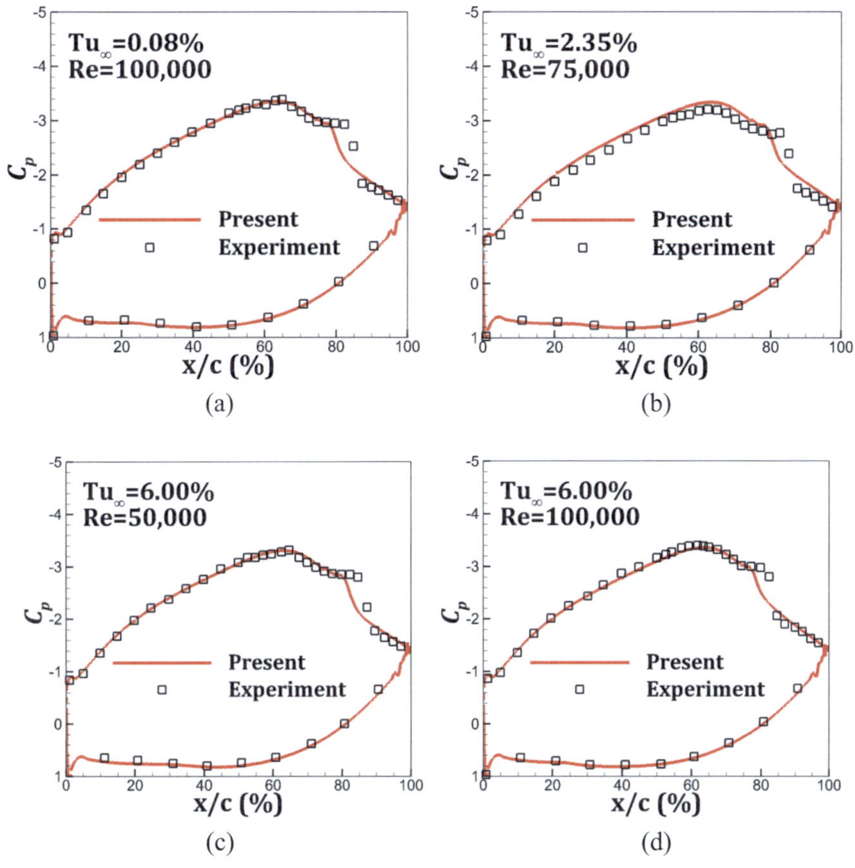

Fig. 3.27 Comparison of surface friction coefficient on the Pak-B low-pressure turbine blade

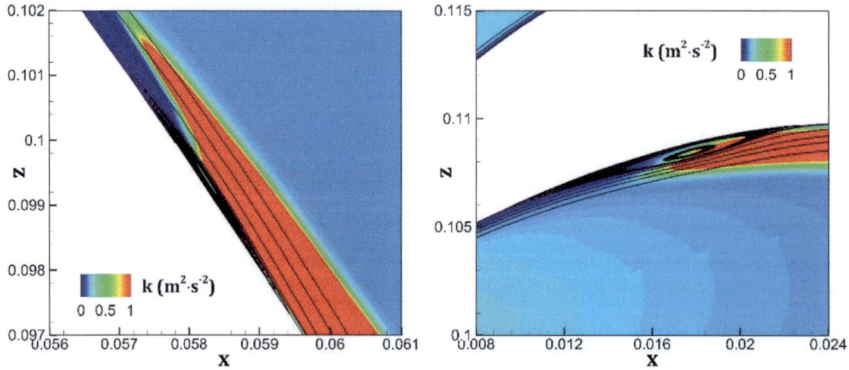

(a) Suction side separation bubble (b) Pressure side separation bubble

Fig. 3.28 Streamline diagram of the Pak-B low-pressure turbine blade at Re = 10,000 and Tu = 6%

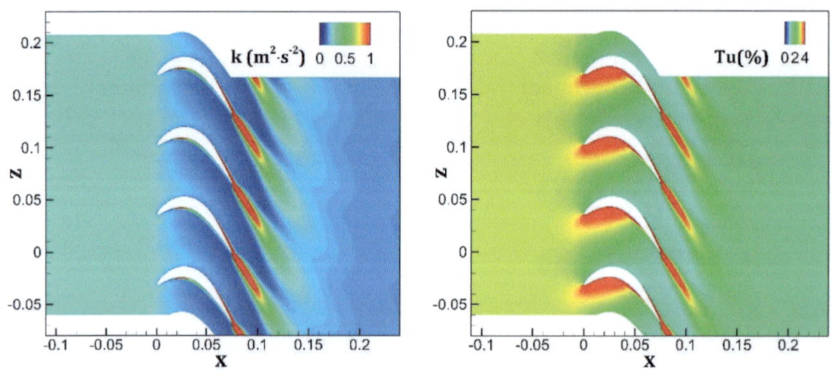

(a) Suction Side Separation Bubble (b) Pressure Side Separation Bubble

Fig. 3.29 Contour plots of turbulent kinetic energy and turbulence intensity for the Pak-B low-pressure turbine cascade at Re = 10,000 and Tu = 6%

3.3.5 V103 Compressor Cascade

In a series of experimental investigations, Hilgenfeld and Pfitzner [184] examined the V103 compressor cascade, which comprises three NACA65 blades. Following this, Zaki et al. [4] employed DNS methods to examine the flow dynamics under varying free-stream turbulence intensities. This section presents simulations of the V103 compressor blades using the recently developed transition model under two specific free-stream turbulence intensity conditions (3.25 and 6.5%) in order to analyze the transition process.

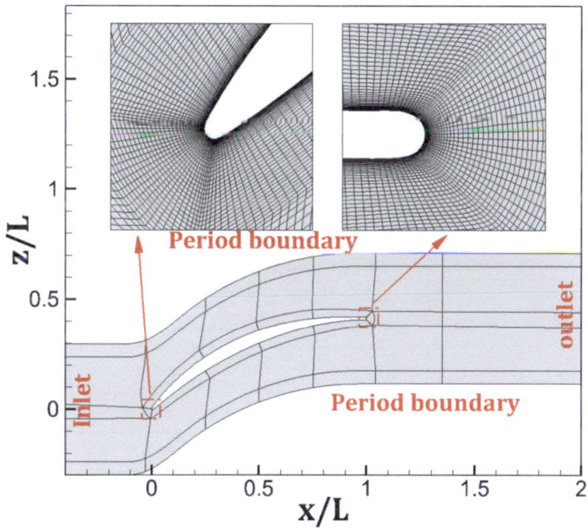

Fig. 3.30 Computational domain and partial mesh distribution for the V103 case

Figure 3.30 illustrates the distribution of the mesh within the computational domain and in the vicinity of the leading and trailing edges of the blade. The inlet boundary is located at $x/L = -0.4$, while the outlet boundary is at $x/L = 2.0$. Moreover, the reference length is the axial chord length L. Periodic boundaries have been applied along the midline of the cascade pitch. The chord length is $L = 203.25$ mm, and the incoming flow velocity is 10 m/s, yielding a Reynolds number of 138,500. The angle between the inlet mean velocity and the horizontal axis is 42°. In order to ensure that the turbulence intensity distribution in the flow field aligns with the DNS data, it is necessary to adjust the inlet viscosity ratio in the course of these simulations. Figure 3.31 illustrates the correlation between the predicted outcomes and the DNS data.

Figures 3.32 and 3.33 present a comparison between the predictions of the surface friction coefficient and pressure coefficient distributions on the blade by the new transition model and DNS data. The figures also include results from the γ [69] and $\gamma - \overline{\mathrm{Re}_{\theta t}}$ [68] transition models. With regard to the suction side, it was observed that as the turbulence intensity increased, the transition location predicted by the new model shifted upstream, which is consistent with the data obtained from the DNS. In both instances, the novel model and the γ model capture the separation bubble on the suction side of the blade; nevertheless, the transition onset and reattachment points are located further upstream compared to the DNS data. The $\gamma - \overline{\mathrm{Re}_{\theta t}}$ model, however, does not predict the formation of a separation bubble on the suction side of the blade. When compared to the DNS data, all three RANS transition models predict a delayed transition location on the pressure side, with the size of the separation bubble being smaller than that observed in the DNS results. Figure 3.34 depicts the turbulent kinetic energy contours for the V103 compressor cascade. As a consequence of the variations in turbulence intensity, the turbulent kinetic energy at the inlet and around the blade is higher for $Tu = 6.5\%$ compared with $Tu = 3.25\%$. Furthermore, higher

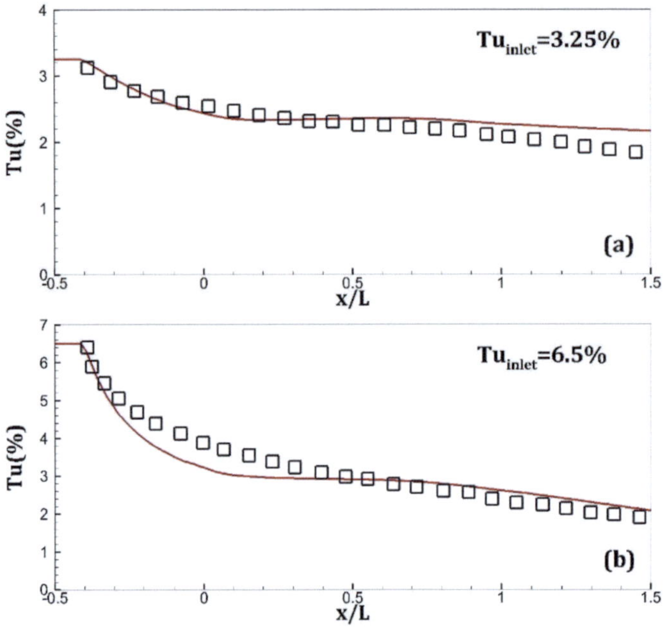

Fig. 3.31 Turbulence intensity distribution along the pitch midline for the V103 case

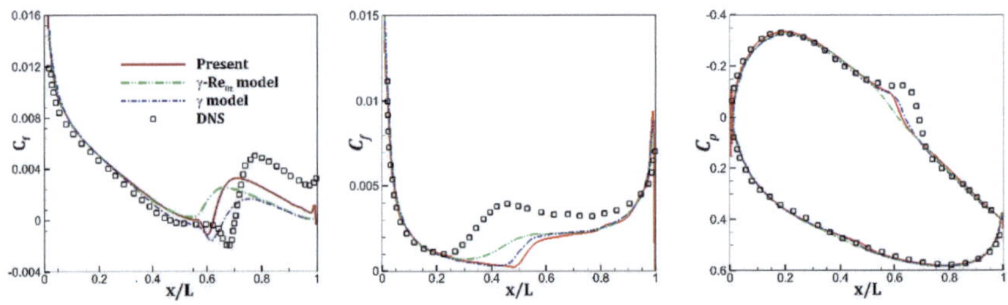

(a) Suction Side Friction Coefficient (b) Pressure Side Friction Coefficient (c) Pressure Coefficient

Fig. 3.32 Distribution of surface friction coefficient and pressure coefficient on V103 compressor blade, $Tu = 3.25\%$

turbulence intensity results in an earlier transition and a shorter separation bubble length. The findings show that the novel model is markedly responsive to alterations in turbulence intensity.

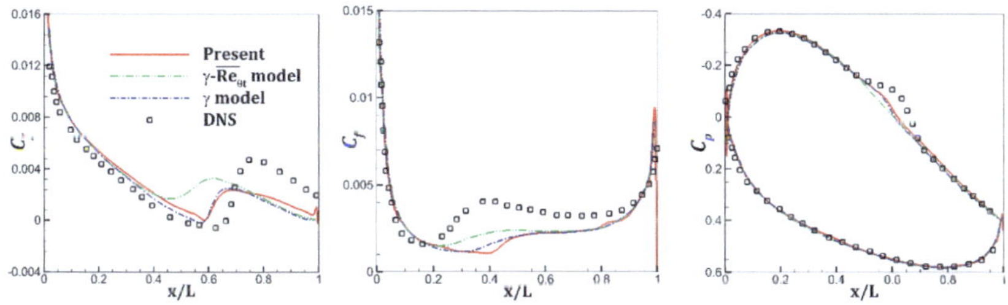

(a) Suction Side Friction Coefficient (b) Pressure Side Friction Coefficient (c) Pressure Coefficient

Fig. 3.33 Distribution of surface friction coefficient and pressure coefficient on V103 compressor blade, $Tu = 6.5\%$

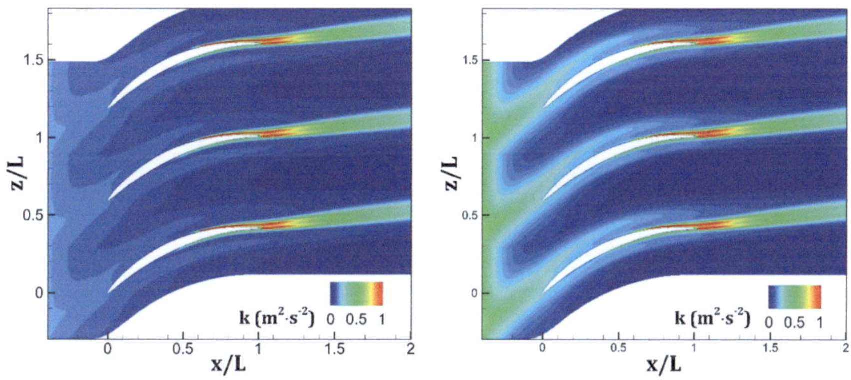

(a) Tu=3.25% (b) Tu=6.5%

Fig. 3.34 Turbulent kinetic energy contour of V103 compressor cascade

3.3.6 NACA0018 Airfoil

To examine the transition process induced by the laminar separation bubble on the upper surface of the NACA0018 airfoil, Istvan and colleagues [167, 185] conducted comprehensive experimental investigations across varying Reynolds numbers and turbulence intensities. This section presents numerical simulations at a chord Reynolds number of 80,000, with free-stream turbulence intensities ranging from 0.006 to 1.99%, to assess the effects of turbulence intensity on the transition behavior of the airfoil's separation bubble. The free-stream angle of attack is maintained at 4°, and the simulation parameters are detailed in Table 3.7 for reference.

The computational domain is configured to approximate the experimental setup, as illustrated in Fig. 3.35. The figure also depicts the grid distribution around the airfoil, and the first grid layer has a y^+ value of less than 0.3.

The pressure coefficient distribution on the upper surface of the airfoil under varying free-stream turbulence intensities is depicted in Fig. 3.36. Despite the suction

Table 3.7 Computational conditions for NACA0018 airfoil

AoA	U_{inlet} (m/s)	Tu_{inlet} (%)	R_T	$\nu(\text{m}^2/\text{s}^{-1})$	Re
4.0	6.0	0.06	8.0	1.5×10^{-5}	80,000
4.0	6.0	0.32	2.8	1.5×10^{-5}	80,000
4.0	6.0	0.51	3.5	1.5×10^{-5}	80,000
4.0	6.0	1.99	27.0	1.5×10^{-5}	80,000

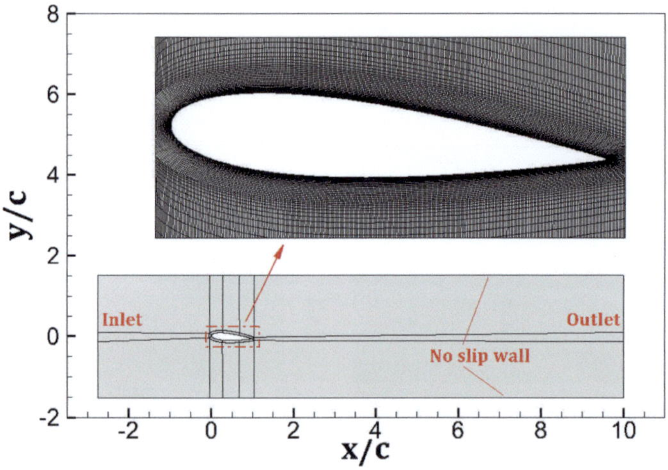

Fig. 3.35 Boundary condition design of the computational domain and local grid distribution for the NACA0018 airfoil

peak being situated in closer proximity to the leading edge, the overall predicted pressure coefficient exhibits a high degree of alignment with the experimental data. Figure 3.37 presents the distribution of turbulent kinetic energy in the vicinity of the laminar separation bubble on the upper surface of the airfoil, as predicted by the transition model. As illustrated in the figure, an increase in free-stream turbulence intensity results in an upstream shift of the transition location, a reduction in the size of the separation bubble, and a corresponding decrease in the peak turbulent kinetic energy downstream of the transition point.

Figure 3.38 depicts the positions of the separation point, transition onset, and flow reattachment point on the upper surface of the airfoil. The results predicted by the transition model are evaluated in comparison with experimental data [185] and the hybrid RANS/LES results reported by Tangermann and Klein [186]. The findings from all three approaches consistently show that with increasing free-stream turbulence intensity, the separation point shifts downstream, while the reattachment point shifts upstream. This leads to a reduction in the size of the separation bubble. The transition locations predicted by both numerical simulation methods are closely

Fig. 3.36 Surface pressure coefficient distribution of the NACA0018 airfoil

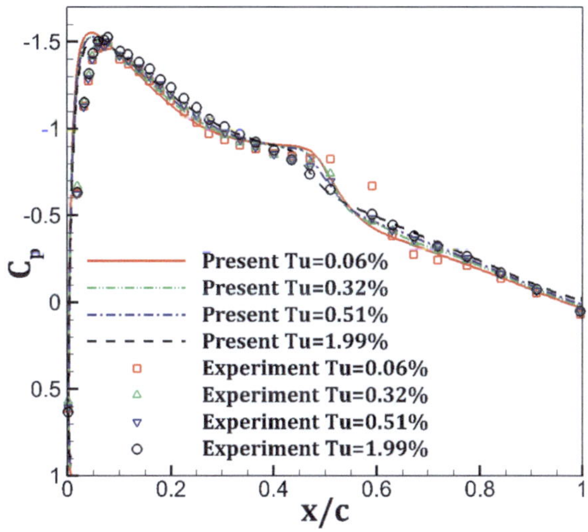

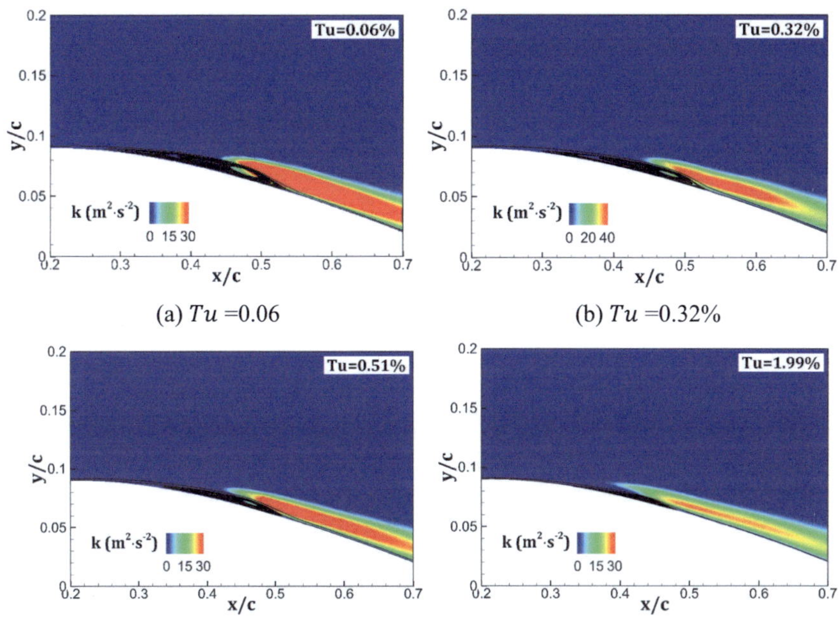

(a) Tu =0.06 (b) Tu =0.32%

(c) Tu =0.51% (d) Tu =1.99%

Fig. 3.37 Contour of turbulent kinetic energy near the laminar separation bubble on the NACA0018 airfoil

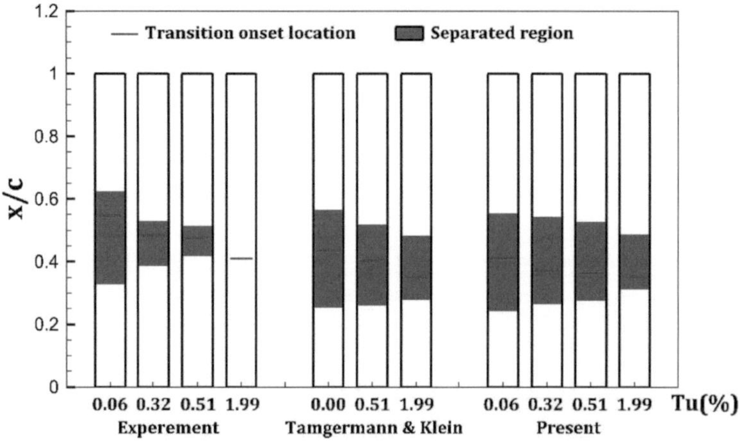

Fig. 3.38 Comparison of transition and separation bubble locations on the NACA0018 airfoil

aligned and are situated upstream of the experimental transition location. A similar trend is observed for the predicted separation point locations.

3.3.7 SD-7003 Airfoil

The Selig-Donovan (SD) 7003 airfoil is a classical low Reynolds number airfoil, often employed to examine the transition processes induced by separation bubbles. At low Reynolds numbers, a laminar separation bubble can form within the boundary layer on the suction surface of the SD-7003 airfoil. Numerous wind tunnel experiments [187–190] and numerical simulations [45, 191–196] have been conducted to analyze the effects of free-stream turbulence intensity and Reynolds number on this laminar separation bubble. In this section, the free-stream turbulence intensities are set to 0.035% and 1.4%, with an angle of attack of 4°. The Reynolds number (Rec), based on the airfoil chord length, ranges from 10,000 to 60,000. The flow conditions are detailed in Table 3.8. Furthermore, Fig. 3.39 illustrates the comprehensive computational domain and the local grid near the airfoil leading edge. The computational grid comprises 781 points arranged circumferentially around the airfoil, with the y^+ value of the first layer grid in the wall-normal direction being less than 1.0.

The results predicted by the transition model at different Reynolds numbers are compared with the data obtained by Galbraith and Visbal [192] using the ILES method. Figure 3.40 depicts the distributions of the surface friction coefficient and pressure coefficient on the airfoil. With increasing Reynolds numbers, the separation and reattachment points on the suction surface boundary layer progressively move upstream, which in turn reduces the size of the laminar separation bubble.

The transition model successfully captures this trend. What's more, the pressure coefficient predictions of the airfoil from the transition model are in good agreement

Table 3.8 Computational conditions for SD-7003 airfoil

AoA	Ma_{inlet} (m/s)	Tu_{inlet} (%)	R_T	L_t/c	$\nu(\text{m}^2/\text{s}^{-1})$	Re_c
4.0	0.1	0.035	0.35	0.90	1.5×10^{-5}	10,000
4.0	0.1	0.035	0.35	0.23	1.5×10^{-5}	40,000
4.0	0.1	0.035	0.35	0.15	1.5×10^{-5}	60,000
4.0	0.1	0.70	7.0	0.13	1.5×10^{-5}	60,000
4.0	0.1	1.40	14.0	0.13	1.5×10^{-5}	60,000

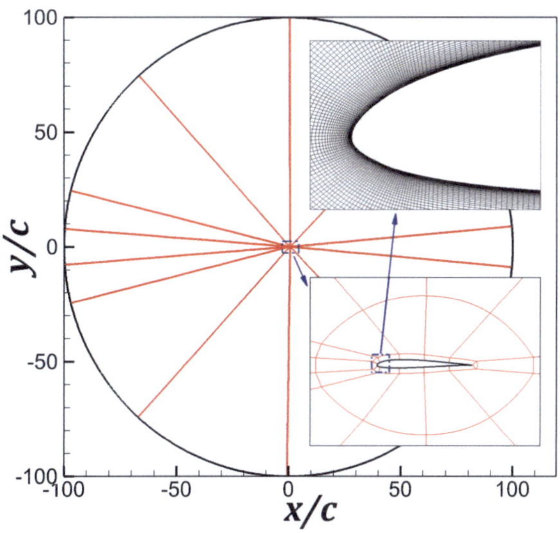

Fig. 3.39 Computational domain and local grid distribution for the SD-7003 airfoil

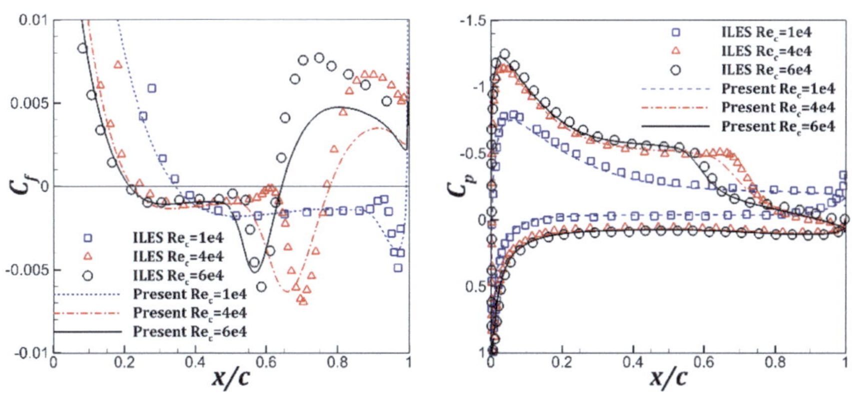

(a) Surface Friction Coefficient (b) Pressure Coefficient

Fig. 3.40 Distribution of surface friction coefficient and pressure coefficient for the SD-7003 airfoil at different reynolds numbers with the ILES results. At $\mathrm{Re}_c = 40{,}000$, the transition model predicts a somewhat smaller laminar separation bubble compared to the ILES data, which contributes to a slightly reduced length of the pressure plateau.

In order to invastigate the influence of the intensity of the free-stream turbulence on the laminar separation bubble of the airfoil, simulations using the transition model were carried out and compared with LES or ILES data from Catalano and Tognaccini [191], Galbraith and Visbal [192], and Breuer [193]. Figures 3.41 and 3.42 illustrate the distributions of the surface friction coefficient and pressure coefficient for the SD-7003 airfoil across different free-stream turbulence intensities. The transition model's predictions of the laminar separation bubble dimensions are in close agreement with the large-eddy simulation data. Nonetheless, downstream of the separation bubble, the surface friction coefficient predicted by the transition model is slightly lower. As the intensity of the free-stream turbulence increases, the size of the laminar separation bubble diminishes and the pressure plateau at the suction surface becomes shorter. The transition model effectively captures these trends, which demonstrates consistency with the large-eddy simulation results.

Figure 3.43 depicts the distribution of streamlines and the effective intermittency factor γ_{eff} in the vicinity of the laminar separation bubble for different free-stream

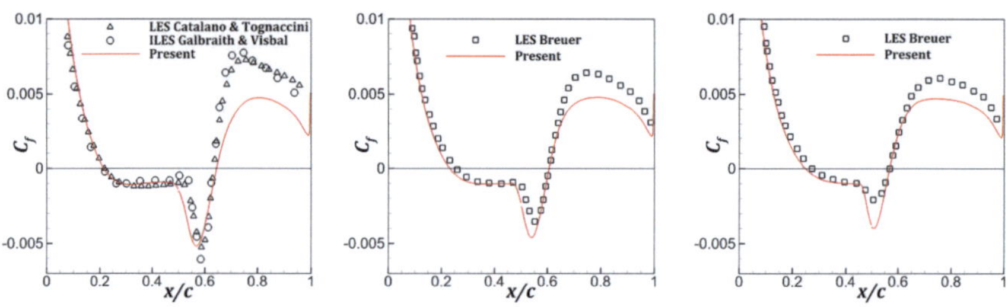

(a) Tu =0.035% (b) Tu =0.7% (c) Tu =1.4%

Fig. 3.41 Comparison of surface friction coefficient for the SD-7003 airfoil under different free-stream turbulence intensities

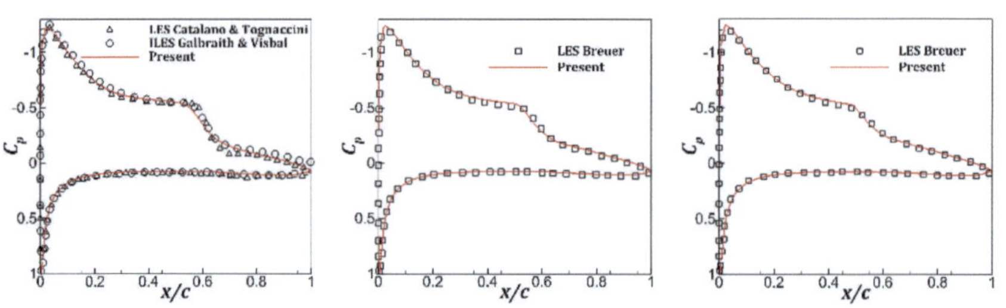

(a) Tu =0.035% (b) Tu =0.7% (c) Tu =1.4%

Fig. 3.42 Comparison of pressure coefficient for the SD-7003 airfoil under different free-stream turbulence intensities

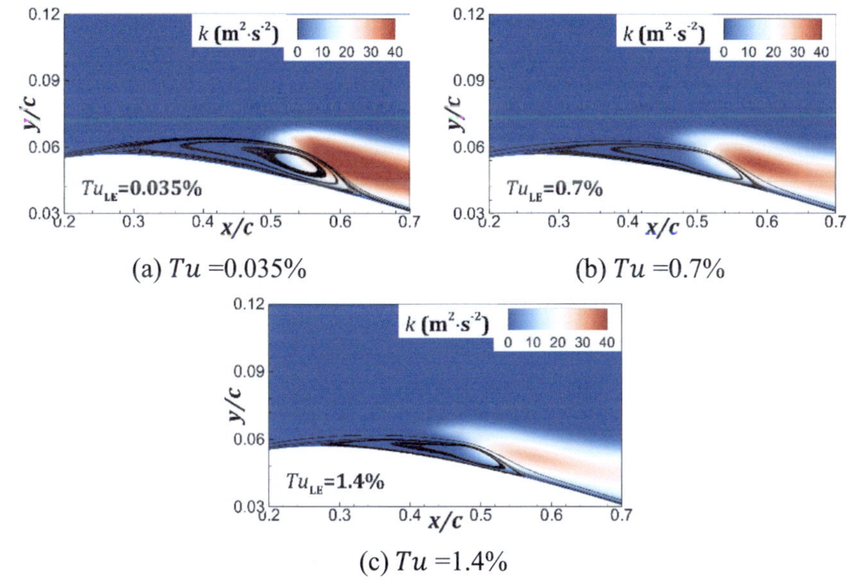

Fig. 3.43 Streamlines and γ_{eff} distribution near the laminar separation bubble of the SD-7003 airfoil under different free-stream turbulence intensities

turbulence intensities. The effective intermittency factor γ_{eff} exceeds 1.0 above the separation bubble, signifying that the transition is initiated within the laminar separation bubble, followed by flow reattachment. With increasing free-stream turbulence intensity, the transition onset migrates closer to the airfoil's leading edge, which facilitates the reattachment of the turbulent boundary layer and contributes to a reduction in the size of the separation bubble.

3.4 Summary

In this chapter, a refined and effective transition-turbulence prediction model is developed within the RANS framework to address bypass transition and transition induced by separation bubbles. This model is founded upon the γ transport equation, offering a simple and efficient method that can be easily incorporated into CFD software. The model's localized variable solution method facilitates its application to large-scale parallel computations. A novel localized indicator is introduced with the objective of capturing the variations in turbulence intensity and pressure gradient. In addition, new criteria for assessing bypass transition are proposed. In case of the transitions induced by laminar separation bubbles, the model employs localized variables to compute the laminar boundary layer shape factor based on the H_L parameter and adjusts the intermittency factor to accurately reflect the effects of free-stream turbulence intensity on the separation bubble size. Validation against multiple classical test cases demonstrates that the model's predictions are in good agreement with experimental and DNS data, indicating its validity and accuracy. Thus, the model is applicable to engineering flow transition predictions, particularly in high-turbulence environments such as those encountered in turbine machinery. Nevertheless, as the model focuses predominantly on bypass and separation bubble transitions, its capabilities for simulating natural transition flows are limited.

Chapter 4
Transition Prediction Model Considering the Effects of Surface Roughness

Abstract Surface roughness effects on aircraft walls significantly influence boundary layer transition locations, thereby profoundly affecting aerodynamic performance and heat transfer characteristics. This chapter proposes a roughness amplification factor transport equation and establishes transition criteria accounting for roughness effects based on wind tunnel experimental data. Coupled with an intermittency factor transport equation, these components form a comprehensive transition prediction model that systematically incorporates wall roughness effects.

4.1 Introduction

Laminar-turbulent transition phenomena are commonly encountered in the flow fields of aircraft, turbomachinery, and gas turbines. Extended operational periods and complex environmental conditions can result in the surface roughening of wings or blades. This phenomenon can be attributed to a number of factors, including dust accumulation, icing, insect debris, chemical erosion, and thermal corrosion [197–201]. The influence of surface roughness on the transition process is significant, with notable implications for the aerodynamic performance, power output, and thermal protection of aircraft, turbomachinery, and gas turbines [202]. In turbomachinery flows, owing to high turbulence intensity, transition primarily occurs through bypass transition and laminar separation bubble-induced transition. Increased surface roughness prompts an upstream shift in the transition onset and a reduction in the length of the laminar separation bubble [203]. In aircraft flows, rough surfaces may diminish or even eliminate the leading-edge separation bubble. Furthermore, a suitably roughened surface can enhance the stall angle of attack and the maximum lift coefficient of the airfoil [204–206]. Given the pivotal role of surface roughness in the evolution of boundary layers and separation bubbles, it is of great importance to accurately predict its effects in numerical simulations of boundary layer transition.

© The Author(s), under exclusive license to Springer Nature Singapore Pte Ltd. 2025
J. Xu et al., *CFD-Compatible RANS/LES Modeling of Transitional and Separated Flows*, https://doi.org/10.1007/978-981-96-6886-1_4

4.2 Development of the Transition Model Considering the Influence of Surface Roughness

For RANS-based transition models, the precise characterization of all disturbances caused by surface roughness remains a formidable challenge. In order to address it, researchers often resort to a more empirical, data-driven method. This method involves utilizing a scalar transport equation linked to the equivalent roughness height, which enables the calculation of variable distributions within the flow field. As a result, the regions affected by surface roughness can be identified. The accurate capture of the scalar distribution within the boundary layer is achieved through the application of boundary conditions on rough surfaces. This scalar is subsequently employed to refine the transition model's criteria, which can quantify the impact of surface roughness on the flow transition process. Inspired by Langel et al. [112, 113], Yang and Xiao [114], and Liu et al. [116], this section develops the transition model presented in the previous chapter. In particular, it incorporates the roughness amplification factor A_r transport equation in order to effectively model the transition induced by roughness in bypass and separation bubble processes. These enhancements extend the predictive scope of the model, culminating in the development of the $\gamma - A_r$ transition model.

4.2.1 *Transport Equation for the Roughness Amplification Factor*

The transport equation for the roughness amplification factor (A_r) is formulated as follows:

$$\frac{\partial(\rho A_r)}{\partial t} + \frac{\partial(\rho u_i A_r)}{\partial x_i} = \frac{\partial}{\partial x_i}\left[\sigma_{Ar}\left(\mu + \frac{\mu_t}{\sigma_{\mu t}}\right)\frac{\partial A_r}{\partial x_i}\right] \tag{4.1}$$

The model coefficients σ_{Ar} and $\sigma_{\mu t}$ are assigned values of 30.0 and 0.4 respectively. This equation determines the distribution of the roughness amplification factor throughout the flow field and effectively conveys the influence of the rough wall surface downstream.

At the wall boundary, the value of the roughness amplification factor is determined by the dimensionless roughness height:

$$A_r|_{\text{wall}} = \frac{100}{1 + \mathrm{e}^{7.15-0.55k_s^+}} - \frac{100}{1 + \mathrm{e}^{7.15}}, \tag{4.2}$$

$$k_s^+ = \frac{u_\tau k_s}{\nu}, \tag{4.3}$$

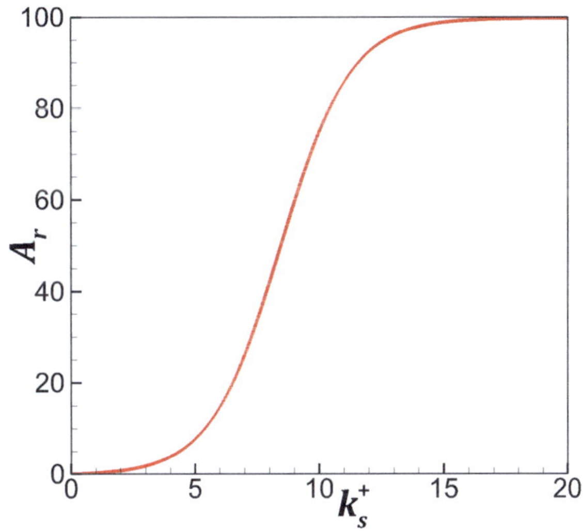

Fig. 4.1 Relationship between the roughness amplification factor and roughness height on the wall

where u_τ is the friction velocity at the wall, and k_s^+ is the dimensionless roughness height.

The function is depicted in Fig. 4.1. For hydraulically smooth walls $\left(k_s^+ < 2.25\right)$ [207], the roughness elements are submerged within the viscous sublayer of the boundary layer, and have a negligible impact on the transition. As the roughness height increases, the value of $A_r|_{\text{wall}}$ on the surface escalates rapidly, and the influence of roughness intensifies. When k_s^+ reaches 20, $A_r|_{\text{wall}}$ attains its maximum value and ceases to increase with further roughness height.

4.2.2 Boundary Condition Modification

Accounting for the effects of rough wall surfaces, it is essential to modify the boundary conditions at the wall. In earlier transition models based on the Wilcox $k - \omega$ model [208] or the SST $k - \omega$ model [130], when surface roughness effects needed to be considered [110, 112, 114, 116], the rough wall boundary conditions proposed by Wilcox [88] were frequently adopted. For rough surfaces, the specific dissipation rate at the wall is defined as follows:

$$\omega_{\text{rough,wall}} = \frac{\rho u_\tau^2 S_R}{\mu} \tag{4.4}$$

The dimensionless function S_R is defined as

$$S_R = \begin{cases} \left(\frac{50}{k_s^+}\right)^2 & \text{if } k_s^+ < 25 \\ \frac{100}{k_s^+} & \text{if } k_s^+ \geq 25 \end{cases} \tag{4.5}$$

Nevertheless, because Wilcox's rough wall correction maintains $k = 0$ and $\mu_t = 0$ at the wall, it fails to precisely predict wall friction under these boundary conditions. In addition, this model necessitates extremely high mesh resolution in the near-wall region [209–211]. Moreover, the eddy viscosity limiter in the $k - \omega$ SST model is designed for smooth surfaces and does not account for the presence of the limiter in the viscous sublayer of fully rough surfaces. Hence, the limiter in the $k - \omega$ SST model is unsuitable for transitional rough or fully rough surfaces. Hellsten and Laine [102] identified this issue and proposed a solution, namely the introduction of a modification to the limiter using an additional function. The function is defined as follows:

$$F_d = 1.0 - \tanh\left[\left(\frac{150\nu}{\omega d^2}\right)^4\right] \tag{4.6}$$

Adding it to the SST limiter:

$$\mu_t = \frac{a_1 \rho k}{\max(a_1 \omega, |\Omega| F_2 F_d)} \tag{4.7}$$

Here, d represents the distance to the nearest wall, the constant a_1 is set to 0.31, and Ω denotes the vorticity. The function F_d equals zero in the near-wall region and one elsewhere. This ensures that F_d suppresses the activation of the $k - \omega$ SST model's limiter on transitional or fully rough surfaces.

The equivalent sand-grain method proposed by Nikuradse [89] for simulating fully rough surfaces results in an increase in μ_t near the wall, which disrupts the viscous sublayer. From the perspective of modeling, the near-wall solution and the turbulent kinetic energy at the wall should be in alignment with the values observed in the logarithmic layer of fully developed flow. Accordingly, as illustrated in Eq. 4.7, μ_t can be calculated from k and ω, which requires assigning a non-zero turbulent kinetic energy value at the wall. Knopp et al. [103] suggested that this wall value should be:

$$k|_{\text{wall}} = \phi_{\text{rl}} k_{\text{rough}},\ k_{\text{rough}} = \frac{u_\tau^2}{\sqrt{\beta_k}},\ \phi_{\text{rl}} = \min\left(1, \frac{k_s^+}{90}\right), \tag{4.8}$$

Among them, k_{rough} is taken from the logarithmic layer value of the rough wall model proposed by Durbin et al. [212], which is only valid under fully rough conditions. Meanwhile, Durbin et al. also assumed that k varies linearly with the roughness height k_s^+. To improve the prediction of transitional rough surfaces, the model was further refined using a coupling function ϕ_{r1}. This refinement allows the boundary condition to provide an appropriate value of k at the wall for smooth, transitionally rough and fully rough surfaces. The model is also suitable for simulating flow transitions. Even when a surface is uniformly covered with rough elements, a viscous sublayer remains in the laminar stagnation region. For hydraulically smooth or transitionally rough surfaces, where $k|_{\text{wall}}$ and μ_t rapidly increase in the laminar region, the coupling function ϕ_{r1} helps prevent premature transition.

The value of the specific dissipation rate ω at the wall is

$$\omega|_{\text{wall}} = \frac{u_\tau}{\kappa d_0 \sqrt{\beta_k}}, \tag{4.9}$$

The d_0 is defined based on the rough surface correction for the SA turbulence model [213], as outlined by Aupoix and Spalart [101].

$$d_0 = \phi_{r2} 0.03 k_s, \tag{4.10}$$

$$\phi_{r2} = \min\left[1, \left(\frac{k_s^+}{30}\right)^{\frac{2}{3}}\right] \min\left[1, \left(\frac{k_s^+}{45}\right)^{\frac{1}{4}}\right] \min\left[1, \left(\frac{k_s^+}{60}\right)^{\frac{1}{4}}\right]. \tag{4.11}$$

In this equation, ϕ_{r2} is calibrated according to the flat plate experimental data from Ligrani and Moffat [207] and demonstrates strong performance on both transitional and fully rough surfaces. The formulation of $\omega|_{\text{wall}}$ ensures that the surface friction coefficient increases as the roughness height increases.

Aupoix [104] noted that the roughness model developed by Knopp et al. tends to produce a slight underestimation of the velocity shift associated with transitional rough surfaces. To address this limitation, Aupoix proposed an alternative roughness model that was based on the experimental data provided by Nikuradse [89] and Colebrook et al. [105]. Nevertheless, when Aupoix's model [104] is applied to transition simulations, the calculation of the effective wall roughness distance introduces considerable disturbances, resulting in an increase in μ_t even within the laminar region [214]. In turn, this model can induce premature transition irrespective of the roughness height or Reynolds number. Consequently, in this study, the roughness model proposed by Knopp et al. [103] has been selected to implement the rough wall boundary conditions.

4.2.3 Coupled Transition Model

When roughness elements are present on the surface of an object, disturbances in the boundary layer are intensified, resulting in an upstream shift of the transition point [215]. What's more, the transition model must be modified to accommodate the effects of wall roughness. In order to accurately reflect the impact of roughness on the transition process, a roughness amplification factor A_r is incorporated into the indicator T_ω, as given by:

$$T_{\omega,\text{R}} = f(A_r) T_\omega, \tag{4.12}$$

$$f(A_r) = 1.0 + C_1 A_r^{C_2}, \tag{4.13}$$

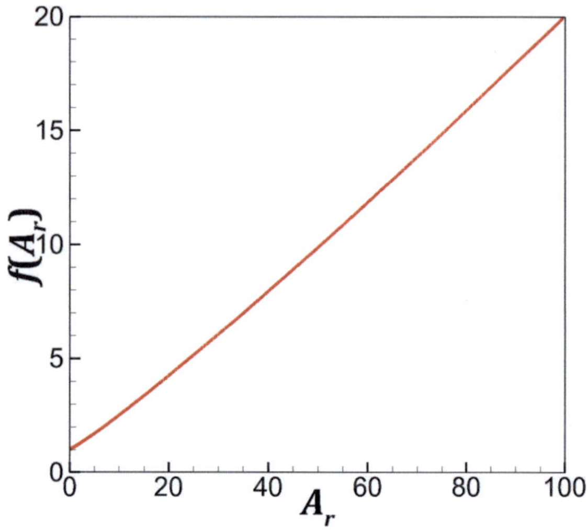

Fig. 4.2 Roughness transition correction function

The model coefficients C_1 and C_2 are 0.12 and 1.1 respectively. The form of the correction function is shown in Fig. 4.2.

Additionally, taking into account the effects of roughness, the critical momentum thickness Reynolds number for transition $\mathrm{Re}_{\mathrm{crit}}$ and the laminar separation bubble-induced transition function f_{modi} are revised as follows:

$$\mathrm{Re}_{\mathrm{crit,R}} = \max\left(\min\left(743.1\mathrm{e}^{-11.61T_{\omega,\mathrm{R}}} + 430.3\mathrm{e}^{-0.7982T_{\omega,\mathrm{R}}}, 1200\right), 40\right), \tag{4.14}$$

$$f_{\mathrm{modi}} = \max(\min(f_{\mathrm{modi1}}, f_{\mathrm{modi2}}), 1.0), \tag{4.15}$$

$$f_{\mathrm{modi1}} = \min\left(1.96e^{93.9T_{\omega,\mathrm{R}}} - 1.05e^{-1959T_{\omega,\mathrm{R}}}, 8.3\right), \tag{4.16}$$

$$f_{\mathrm{modi2}} = \min\left(4.39\mathrm{e}^{0.415T_{\omega,\mathrm{R}}} - 1.74\mathrm{e}^{-29.3T_{\omega,\mathrm{R}}}, 8.3\right). \tag{4.17}$$

In the production term of the transport equation for the intermittency factor γ, the criteria for determining the onset of transition will also be modified.

$$F_{\mathrm{onset1}} = 2.0 \cdot \tanh\left(\left(\frac{\mathrm{Re}_V}{2.193\mathrm{Re}_{\mathrm{crit,R}}}\right)^{30.0}\right). \tag{4.18}$$

As the roughness height k_s^+ increases, the function $f(A_r)$ correspondingly rises, resulting in an increase in the indicator $T_{\omega,\mathrm{R}}$ and thereby intensifying the influence of surface roughness. In due course, a reduction in the critical momentum thickness Reynolds number $\mathrm{Re}_{\mathrm{crit,R}}$ is observed, accompanied by an increase in the transition function f_{modi}, an upstream shift in the transition onset, and a subsequent decrease in the size of the laminar separation bubble.

The other functional relationships within the roughness-induced transition model, along with the turbulence model modifications, stay consistent with the transition model detailed in the previous chapter and will not be reiterated here.

4.3 Case Validation

In this section, the predictive capability and simulation accuracy of the $\gamma - A_r$ transition model are assessed by examining the impact of roughness on transition across various cases, including a flat plate with zero pressure gradient, a flat plate with a pressure gradient, a rough flat plate experiencing boundary layer separation, the NREL-S814 airfoil, and high- and low-pressure turbine blades.

4.3.1 *Rough Flat Plate with Attached Flow*

This section employs Feindt's flat plate flow experiments for validation. The experiments conducted by Feindt establish the relationship between the roughness height and the transition onset position under varying pressure gradients. In the numerical simulations, the inlet height (h_0) and plate length (L) are maintained at a constant value, while the linear pressure gradient is varied by modifying the geometry of the external boundary. The upper boundary profile of the plate with a non-zero pressure gradient is defined by the function $h(x)$, which is expressed as follows:

$$h(x) = \sqrt{\frac{h_0^2}{\sqrt{1.0 - PG}}},\ PG = \frac{p_1(x) - p_0}{q_0} = \pm 4.0x \tag{4.19}$$

In this equation, the inlet height h_0 is set to 0.1495 L, with p_0 and q_0 denoting the static and dynamic pressures at the inlet respectively, while p_1 represents the static pressure along the plate. The plate length is 1 m, and the pressure gradient parameter PG is adjusted to modify the geometry of the upper boundary, aligning with Feindt's non-zero pressure gradient experimental conditions. This function yields two different pressure gradient configurations: a favorable pressure gradient (FPG) plate and an adverse pressure gradient (APG) plate. Together with a zero pressure gradient (ZPG) plate, these form the three cases examined in this section, with the grid distributions illustrated in Fig. 4.3.

The inlet velocity of the flat plate flow field is 20 m/s, and the turbulence intensity of free-stream is 1.0%. In addition, the Reynolds number based on the plate length is 1.3×10^6. Numerical simulations were carried out using the new model for flat plates with eight distinct roughness heights, ranging from 0 to 275 μm, as detailed in Table 4.1. For the smooth surface case, the roughness height k_s was set to 0 μm.

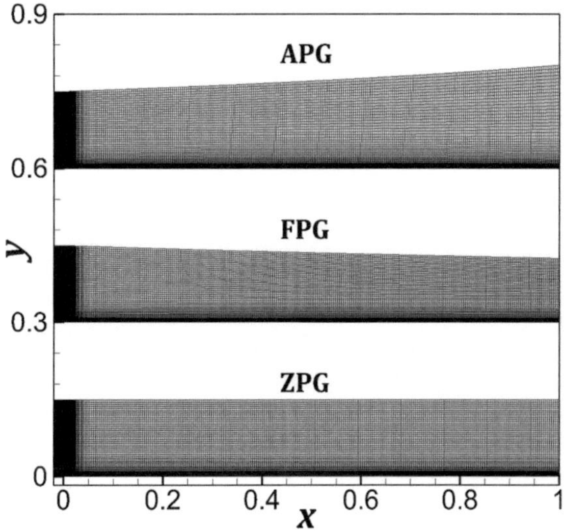

Fig. 4.3 Geometrical configuration and grid distribution of the flat plate

Table 4.1 Boundary conditions for flat plate simulations

Re	$U_{\infty}(\mathrm{m/s})$	$k_s(\mu m)$	$Tu(\%)$
1.3×10^6	20	0, 40, 80, 120, 160, 190, 220, 275	1.0

In order to guarantee the precision of the recently-developed roughness-induced transition model, a grid convergence study was undertaken. This study was performed on a zero-pressure-gradient flat plate using three different grid resolutions, as detailed in Table 4.2. For all three grids, the dimensionless height $\left(y^{+}\right)$ of the first cell layer was kept below 0.5. The results of the simulations, presented in Fig. 4.4, demonstrate the efficacy of these grids in obtaining accurate skin friction coefficients. The transition locations predicted by the "Fine" and "Medium" grids are nearly identical, while the transition location predicted by the "Coarse" grid is slightly upstream. This suggests that the transition location on the flat plate remains stable as the grid is refined. In the end, the "Medium" grid was selected for further simulations, with its distribution shown in Fig. 4.3.

Figure 4.5 illustrates the influence of surface roughness on the friction coefficient $\left(C_f\right)$ and the transition onset position (Re_{xt}) on a zero-pressure-gradient flat plate. In this figure, $\mathrm{Re}_{k_s} = \rho U_{\infty} k_s / \mu$ represents the roughness Reynolds number. The figure

Table 4.2 Parameters of different grids used in the grid convergence analysis

	Coarse	Medium	Fine
Number of streamwise nodes	233	321	429
Number of wall-normal nodes	77	97	133
Growth rate in boundary layer	1.15	1.10	1.05

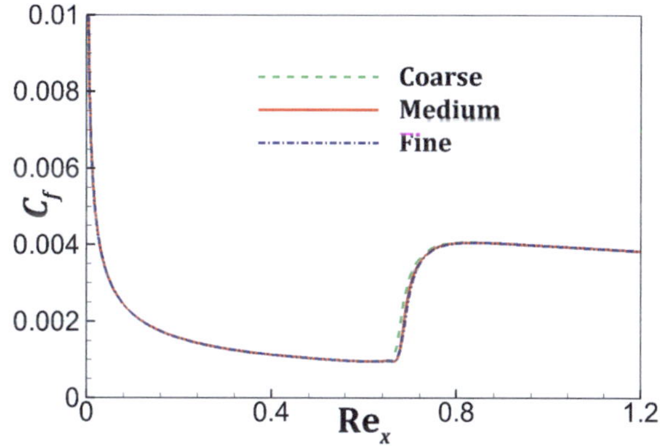

Fig. 4.4 Distribution of surface friction coefficient on a zero-pressure-gradient flat plate for different grids

shows that as the roughness height increases, the transition point moves progressively upstream on the flat plate. In fully turbulent regions, the friction coefficient for the rough surface is higher than that for the smooth surface. The transition onset position is identified as the point where C_f starts to increase sharply. The figure also compares the model's predicted transition onset with experimental data and the numerical results of Dassler et al. [111] and Lagnel et al. [113].

In practical flow scenarios, when the roughness height is very small, giving rise to a plate surface that is nearly hydraulically smooth, the roughness elements are embedded within the viscous sublayer of the boundary layer, exerting minimal influence on the flow. As a result, there is only a slight change in the transition onset position when transitioning from a smooth surface to minimal roughness. Nonetheless, as the roughness height increases, the transition location shifts with greater rapidity. Once the roughness height reaches a sufficiently large value, the transition point approaches the leading edge, and the changes in transition onset position

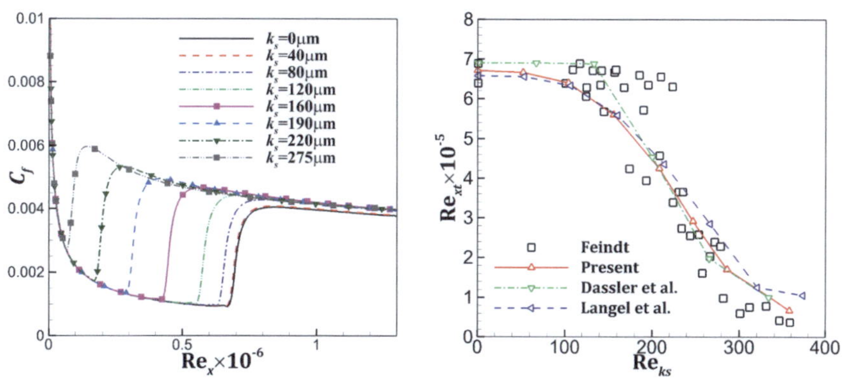

(a) Distribution of Surface Friction Coefficient

(b) Variation of Transition Onset Position with Roughness Height

Fig. 4.5 Effect of roughness on transition flow over a zero-pressure-gradient flat plate become more gradual. The new model effectively captures this behavior, and its predicted transition location variation aligns well with the results from Langel et al. and Dassler et al.

At three different roughness heights ($k_s = 0,\ 40$ and $190\ \mu m$), the distributions of the roughness growth factor (A_r) and intermittency factor (γ) in the zero-pressure-gradient flat plate flow field are illustrated in Figs. 4.6 and 4.7. When $k_s = 40\ \mu m$, the plate surface is smooth, contributing to an A_r value of zero throughout the flow field. At $k_s = 40\ \mu m$, A_r remains very small, close to zero, and there is minimal change in the transition onset position compared to the smooth plate. In contrast, at $k_s = 190\ \mu m$, the wall value of A_r is significantly higher, and this roughness growth factor is transported and diffused into the boundary layer near the wall through the transport equation, eventually dissipating as it moves away from the wall. While A_r exerts an influence over a broader area within the turbulent boundary layer, its principal effect is observed within the laminar boundary layer. This results in a reduction in the critical momentum thickness Reynolds number, an increase in γ, and a subsequent upstream shift in the transition location.

To further investigate the effect of roughness on transition, the distributions of A_r, $T_{\omega,\mathrm{R}}$, $\mathrm{Re}_{\mathrm{crit}}$, and k within the boundary layer at the streamwise position $x/L = 0.2$ are extracted, as shown in Fig. 4.8. For the three flat plates with roughness heights of $k_s = 0,\ 40$ and $190\ \mu m$, the transition occurs downstream of $x/L = 0.2$. On the smooth plate, the A_r value remains at zero. For the plate with $k_s = 40\ \mu m$, the A_r value

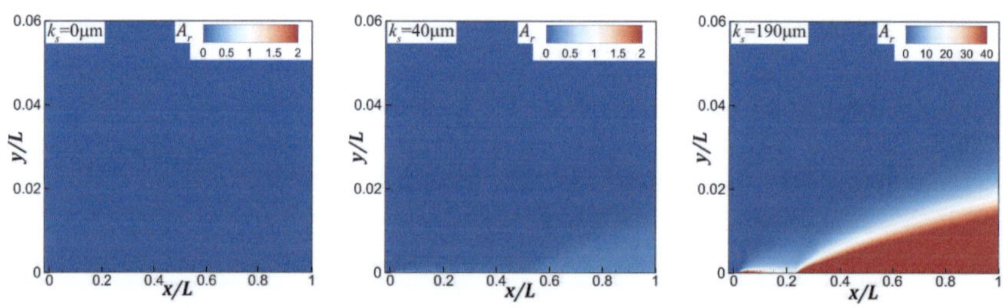

(a) k_s=0 μm (b) k_s=40 μm (c) k_s=190 μm

Fig. 4.6 Distribution of A_r in the flow field of a zero-pressure-gradient flat plate

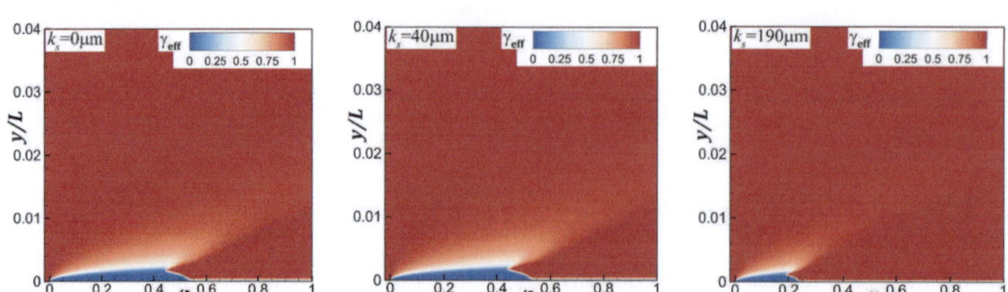

(a) k_s=0 μm (b) k_s=40 μm (c) k_s=190 μm

Fig. 4.7 Distribution of γ in the flow field of a zero-pressure-gradient flat plate

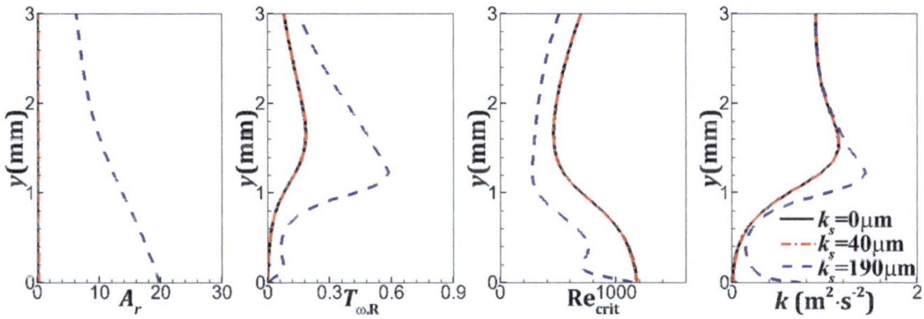

Fig. 4.8 Distributions of A_r, $T_{\omega,\mathrm{R}}$, $\mathrm{Re}_{\mathrm{crit,R}}$, and k within the boundary layer at $x/L = 0.2$ for different rough flat plates

within the boundary layer is extremely low, which results in little to no change in $\mathrm{Re}_{\mathrm{crit}}$ and k compared to the smooth plate. Nevertheless, when the roughness height is 190 μm, A_r increases to 4 within the boundary layer and gradually decreases with distance from the wall. The reduction in $\mathrm{Re}_{\mathrm{crit}}$ causes a shift of the transition location upstream. Furthermore, due to the roughness boundary correction proposed by Knopp et al., the turbulence kinetic energy k near the wall remains non-zero for larger roughness heights, bringing about higher k values in the boundary layer near the wall, which differs from the distributions observed on the other two plates.

Figures 4.9 and 4.10 respectively depict the impact of surface roughness on the transition process for flat plates under favorable-pressure-gradient (FPG) and adverse-pressure-gradient (APG) conditions. As with the zero-pressure-gradient plate, experimental data and numerical results from Lagnel et al. [113] and Dassler et al. [111] are provided for comparison. The variation in transition onset location with increasing roughness height on FPG and APG plates follows a trend similar to that of the zero-pressure-gradient plate. Specifically, the favorable pressure gradient shifts the transition downstream, while the adverse pressure gradient shifts it upstream. The new roughness transition model effectively captures these behaviors, demonstrating strong alignment with experimental data and showing good agreement with the numerical results reported in the literature.

4.3.2 Rough Flat Plate with Boundary Layer Separation

Jeong and Song [216] performed research on the separated boundary layer over a flat plate in the low-speed, open-circuit wind tunnel at Seoul National University (SNU). To simulate the suction-side pressure gradient of the high-lift, low-pressure turbine T106A airfoil [217], a specially contoured wall was installed in the test section. The flat plate was centrally positioned within the test section, with wall surfaces placed on both the upper and lower sides to maintain a 0° angle of attack for the incoming flow. The $\gamma - A_r$ transition model was applied to simulate the flow field, with the computational domain and mesh configuration shown in Fig. 4.11. The upper

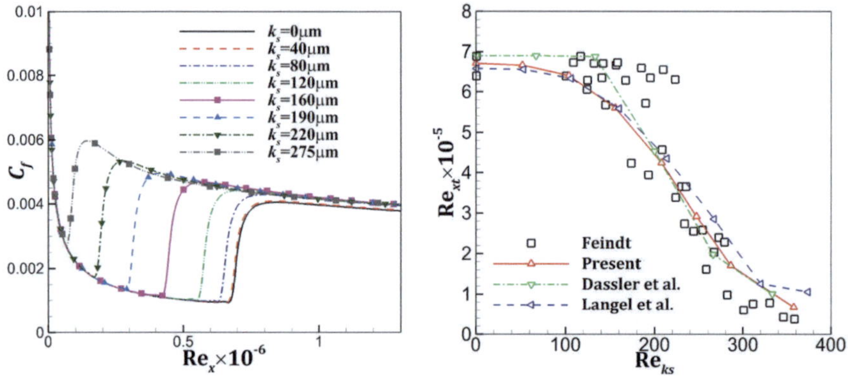

(a) Surface Friction Coefficient Distribution

(b) Variation of Transition Onset Position with Roughness Height

Fig. 4.9 Effect of roughness on the transition flow over a favorable-pressure-gradient flat plate

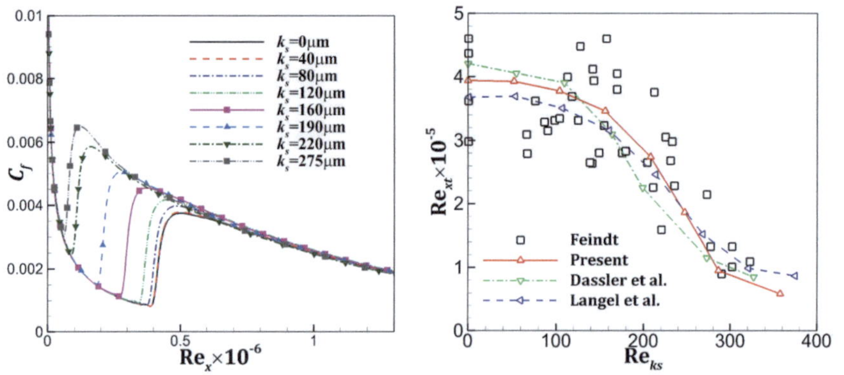

(a) Surface Friction Coefficient Distribution

(b) Variation of Transition Onset Position with Roughness Height

Fig. 4.10 Effect of roughness on the transition flow over an adverse-pressure-gradient flat plate

boundary of the flow field was aligned with the test section contour reported by Jeong and Song and was treated as a slip wall. The grid on the flat plate consisted of 545 points in the streamwise direction and 93 points normal to the wall, with a normal grid growth rate of 1.1. The first layer of the grid near the flat plate had a y^+ value of less than 0.6. In the experiment, the flat plate had a length of $c = 546$ mm, and the flow velocity measured at the test section exit was $U_0 = 3.45$ m/s, resulting in a flow Reynolds number Re_c of 1.2×10^6. Table 4.3 summarizes the computational conditions for this case. The experiment involved three flat plates with different roughness heights: $k_s = 1.27,\ 131.7,\ 211.5\ \mu m$. The initial plate, given its minimal roughness height, was treated as hydraulically smooth and designated "Smooth."

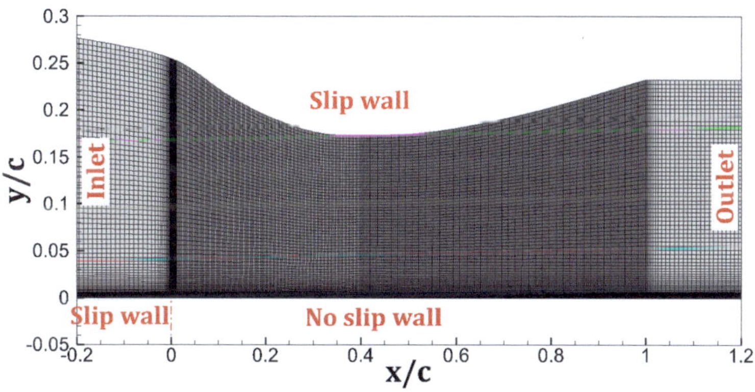

Fig. 4.11 Computational domain and mesh distribution for the separated flat plate

Table 4.3 Computational conditions for the separated flat plate flow

Case	$k_s(\mu m)$	U_0	Re_c	Tu_{in} (%)	R_T
Smooth	1.27	3.45	1.2×10^5	3.2	5.0
Rough1	131.7	3.45	1.2×10^5	3.2	5.0
Rough2	221.5	3.45	1.2×10^5	3.2	5.0

The turbulence intensity Tu_{in} at the leading edge of the flat plate was determined to be 3.2%.

Figure 4.12 presents a comparison between the static pressure coefficient distribution obtained from simulations of the smooth flat plate and the experimental results reported by Jeong and Song [216] and Stieger [217]. The comparison reveals a robust correlation between the numerical predictions and the experimental measurements. What's more, Fig. 4.13 presents the surface friction coefficient distributions for the flat plate across a range of roughness heights. The laminar separation and turbulent reattachment points, as measured experimentally by Jeong and Song [216], are also shown. The simulations indicate a slight delay in the laminar separation point relative to the experimental data, with separation locations for different roughness heights remaining consistent. As roughness height increases, the turbulent reattachment point moves progressively upstream. In general, the numerical predictions of the separation bubble size closely match the experimental results.

Figure 4.14 shows the velocity and streamline distributions across flat plates with varying roughness heights. The "Present" label denotes the results from the simulations, while "Exp." refers to the experimental measurements. In the initial half of the plate length, the laminar boundary layer forms under the influence of the favorable pressure gradient. Around 60% of the plate length, the adverse pressure gradient causes the laminar boundary layer to separate, which leads to the formation of a laminar separation bubble followed by reattachment. With increasing roughness

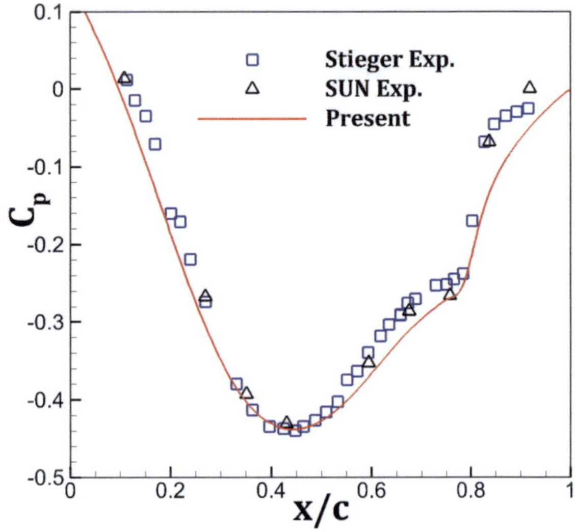

Fig. 4.12 Comparison of static pressure coefficient distribution for the smooth flat plate

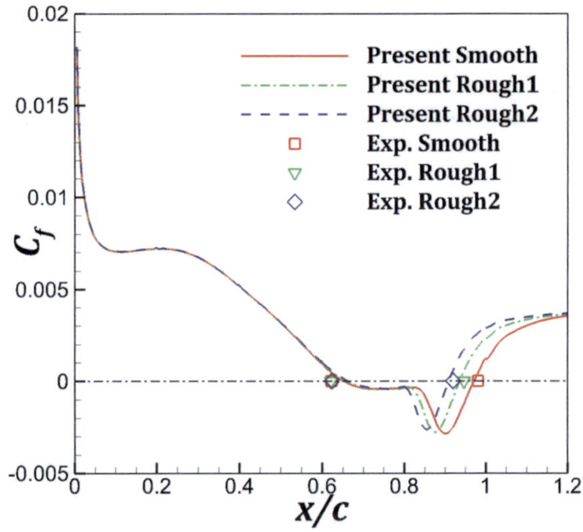

Fig. 4.13 Comparison of surface friction coefficient distributions for flat plates with different roughness heights

height, the separation onset position remains relatively unchanged, although both the height and length of the separation bubble are notably reduced.

The transition model effectively captures this behavior, and the predicted length and height of the separation bubble for various roughness heights closely match the experimental findings.

Figure 4.15 depicts the distribution of turbulent kinetic energy near the laminar separation bubble on various flat plates. The transition to turbulence is initiated within the laminar separation bubble, leading to a gradual increase in turbulent kinetic energy as the boundary layer evolves into a turbulent state. The peak turbulent kinetic

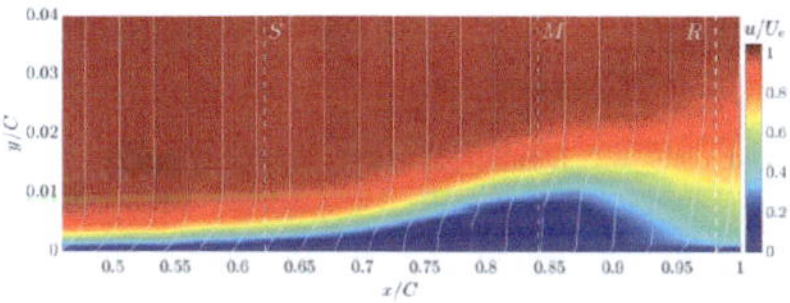

(a) Exp. Smooth

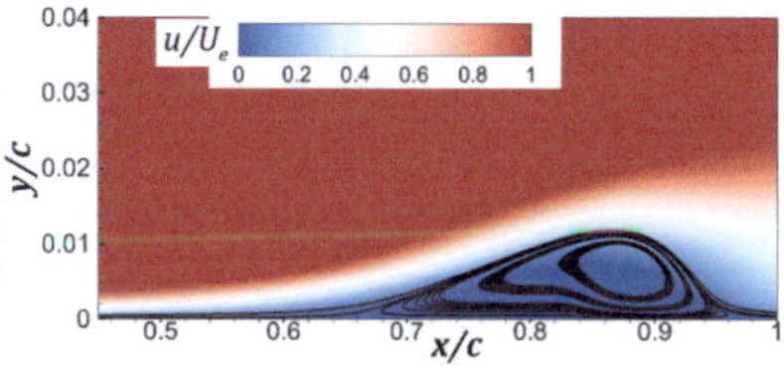

(b) Present Smooth

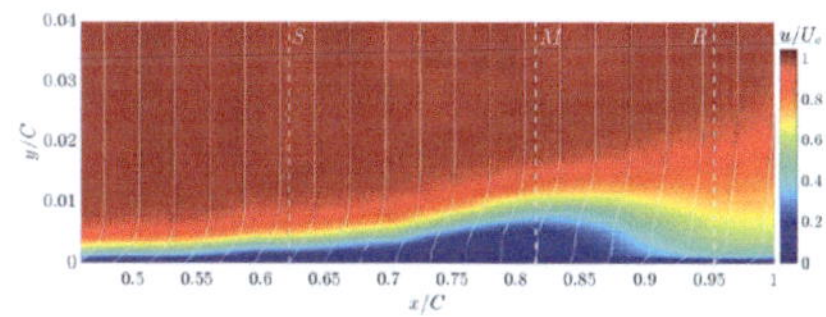

(c) Exp. Rough1

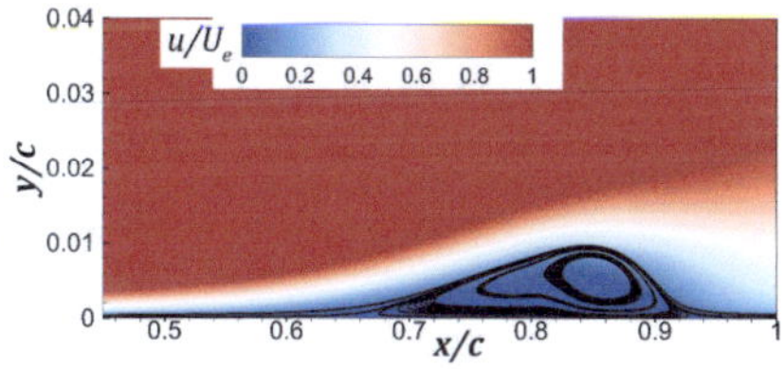

(d) Present Rough1

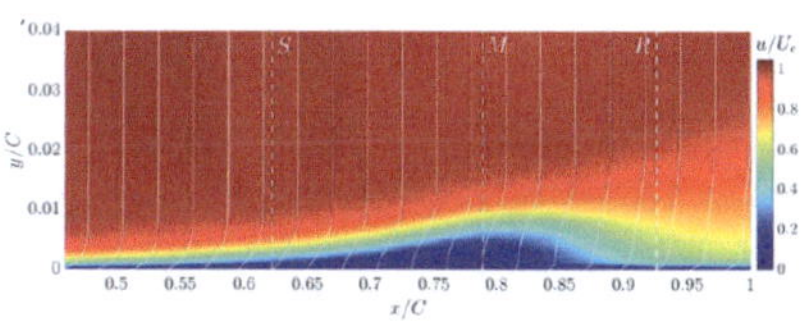

(e) Exp. Rough2

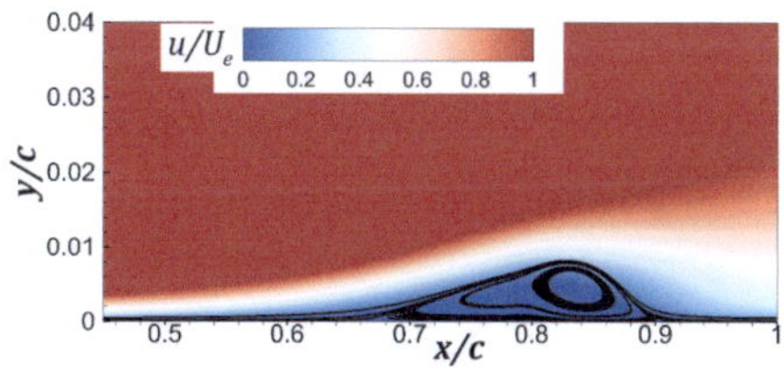

(f) Present Rough2

Fig. 4.14 Velocity and streamline distribution on flat plates with different roughness heights

energy is observed downstream of the separation bubble's maximum height. As the roughness height of the plate increases, the transition location shifts progressively upstream, causing the onset of increased turbulent kinetic energy to also advance upstream. It also results in a reduction in the size of the separation bubble and a corresponding decrease in the maximum turbulent kinetic energy within the flow field.

Figure 4.16 presents the velocity profiles at the transition onset location and at $x/c = 0.981$ for various roughness heights. As a result of the experimental limitations in measuring reverse flow velocity, some discrepancies exist between the predicted reverse flow within the separation bubble and the experimental data. Nevertheless, the predicted velocity profiles align closely with the measured data in other regions [216]. At the transition onset, increasing roughness height leads to a thinner boundary layer and a reduction in reverse flow velocity. In addition, as surface roughness increases, the reattachment point shifts upstream, resulting in a thinner boundary layer at $x/c = 0.981$, with the velocity profile increasingly resembling that of a fully developed turbulent boundary layer. This demonstrates that surface roughness advances the transition and enhances turbulence. Overall, the $\gamma - A_r$ transition model effectively captures the influence of roughness on the transition process induced by the laminar separation bubble.

Fig. 4.15 Turbulent kinetic energy distribution near separation bubbles on plates with different roughness heights

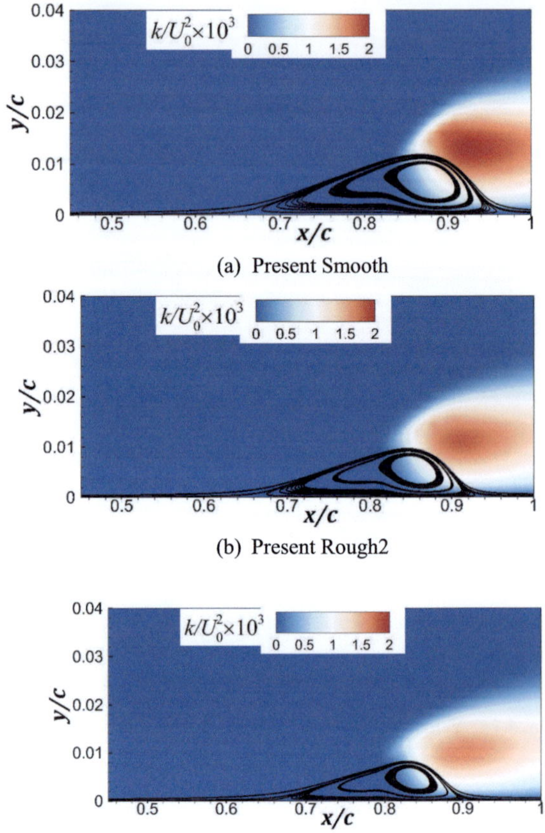

(a) Present Smooth

(b) Present Rough2

(c) Present Rough2

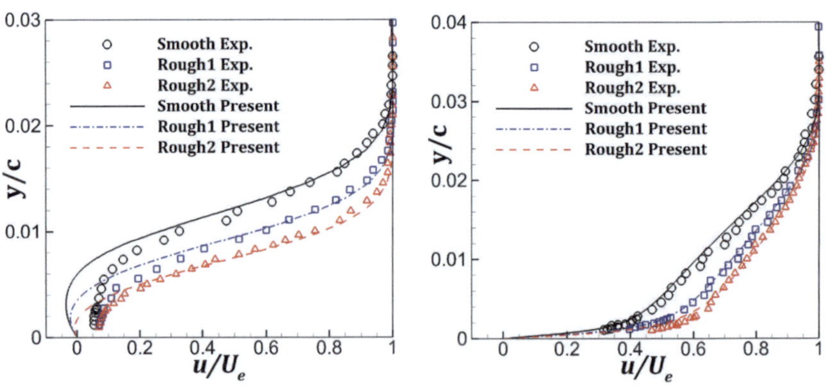

(a) Velocity Profile at the Transition Onset Location

(b) Velocity Profile at $x/c = 0.981$

Fig. 4.16 Comparison of velocity profiles on flat plates with different roughness heights

4.3.3 Rough Leading Edge NREL-S814 Airfoil

In a comprehensive experimental investigation, Janiszewska et al. [218] examined the NREL-S814 airfoil in the subsonic wind tunnel at The Ohio State University, with a particular emphasis on the impact of leading-edge contamination on the airfoil's aerodynamic performance. In this section, the advanced transition model will be used to perform numerical simulations of the airfoil flow, with the results subjected to critical comparison with the experimental data.

In the experiment, the chord length of the model was $c = 457$ mm, with a sandpaper roughness strip applied at the leading edge to simulate the contamination of the airfoil. The roughness elements had a height of $k_s/c = 0.0019$. This roughened area spanned 102 mm, covering both the upper and lower surfaces at the leading edge in a uniform manner. In order to conduct the numerical simulations, the freestream Mach number was set to 0.1, with a chord-based Reynolds number of 1.0×10^6 and a freestream turbulence intensity of 0.5%. The $\gamma - A_r$ transition model was employed for both the smooth airfoil and the roughened leading-edge airfoil, with the roughness height k_s set to zero for the smooth case. Figure 4.17 illustrates the computational domain for the NREL-S814 airfoil, together with the distribution of the mesh near the airfoil. The mesh comprises 416 grid points along the airfoil surface and 181 points in the normal direction. The first layer mesh achieves a y^+ value of less than 0.5.

Figure 4.18 presents a comparative analysis of the aerodynamic coefficients between the smooth and roughened airfoils. At identical angles of attack, the roughened airfoil exhibits a lower lift coefficient (C_L) than its smooth counterpart. In addition, for a given lift coefficient, the drag coefficient (C_D) is substantially higher for the roughened airfoil. The transition model's predictions for the smooth airfoil's lift and drag coefficients align closely with the experimental data, thereby demonstrating high accuracy. For the roughened airfoil, the model's lift coefficient predictions are in close agreement with experimental results at small angles of attack ranging from

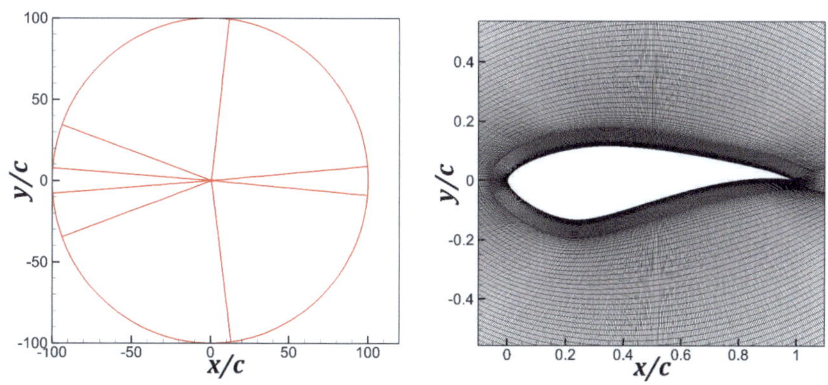

(a) Mesh Topology (b) Grid Distribution Near the Airfoil

Fig. 4.17 Computational domain and grid distribution near the NREL-S814 airfoil

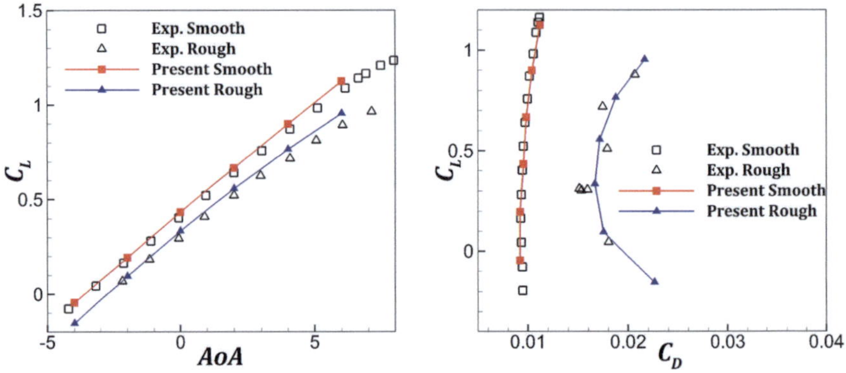

(a) Variation of Lift Coefficient with Angle of Attack (b) Polar Curve

Fig. 4.18 Lift coefficient curve and polar curve for the NREL-S814 airfoil

− 4° to 2°. Nevertheless, at higher angles of 4° and 6°, the model produces a slightly higher lift coefficient than is observed in the experiments. This suggests that the transition model may predict a longer laminar flow region than is actually the case.

Figure 4.19 illustrates the pressure coefficient distribution for the NREL-S814 airfoil at a 0° angle of attack. The introduction of roughness on the leading edge has been observed to result in a significant reduction in the pressure peak on the upper surface, from − 0.92 to − 1.35. Concurrently, the pressure coefficient near the peak on the lower surface has been found to increase by approximately 0.05. This modification results in a reduction of the pressure differential between the upper and lower surfaces, which in turn causes a decrease in the lift coefficient from 0.431 to 0.327 in comparison to the smooth airfoil.

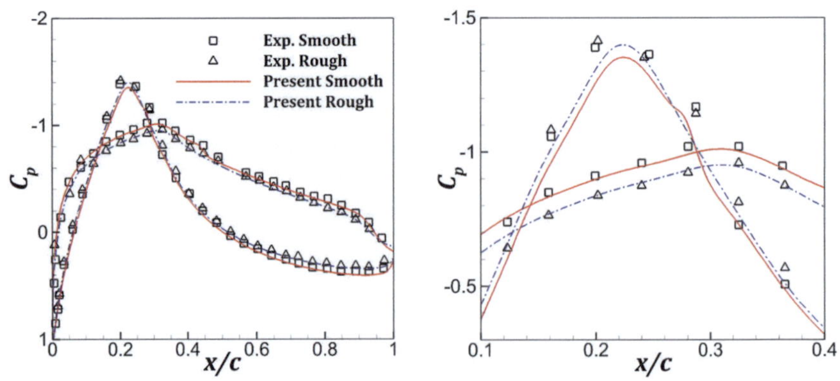

(a) Overall Distribution (b) Local Distribution

Fig. 4.19 Pressure coefficient distribution for the NREL-S814 airfoil at 0° angle of attack

Figure 4.20 depicts the pressure coefficient distribution of the NREL-S814 airfoil at a 6° angle of attack. At this angle, the influence of the roughened leading edge on the pressure distribution is more pronounced than at 0°. The presence of a rough surface has a marked effect on the pressure coefficient, resulting in a notable reduction on the upper surface and an increase on the lower surface. This leads to a decline in the lift coefficient, from 1.12 to 0.95. The magnified view reveals a sharp variation in the pressure coefficient around 35% of the chord length on the smooth airfoil, which indicates the presence of a laminar separation bubble and subsequent transition. Figure 4.21 presents the distribution of the surface friction coefficient and effective intermittent factor for the smooth NREL-S814 airfoil at a 6° angle of attack. Consistent with the pressure coefficient distribution shown in Fig. 4.20, both the upper and lower surface friction coefficients of the smooth airfoil exhibit regions with values below zero near 35% of the chord length, indicating the formation of a separation bubble within the boundary layer. This is corroborated by the effective intermittent factor distribution, which substantiates the hypothesis that the laminar separation bubble is the causal factor in the transition from laminar to turbulent flow. In the case of the roughened airfoil, the presence of the leading-edge results in a change in the transition mechanism, whereby the transition is no longer laminar separation bubble-induced, but rather bypass transition-related, occurring in proximity to the leading edge. This results in a more gradual pressure coefficient distribution, with less abrupt changes, in comparison to the smooth airfoil. Figure 4.22 shows the distribution of the roughness growth factor A_r within the flow field, with higher A_r values observed in proximity to the rough surface, which subsequently decrease downstream. The findings demonstrate that the roughness transition model accurately reflects the impact of leading-edge roughness on the pressure coefficient distribution on both the suction and pressure surfaces and effectively simulates the effect of surface roughness on flow transition patterns.

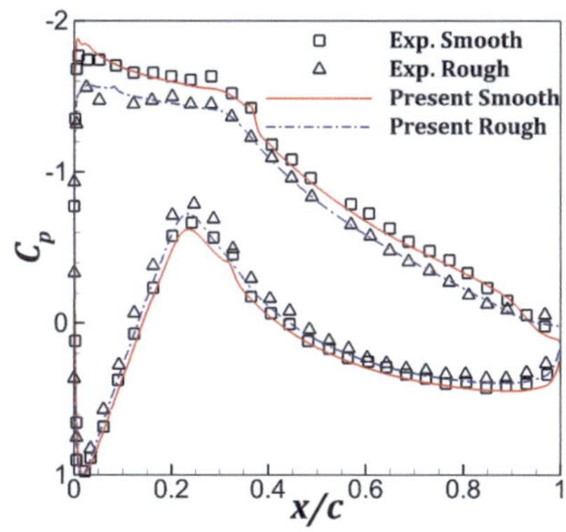

(a) Overall Distribution

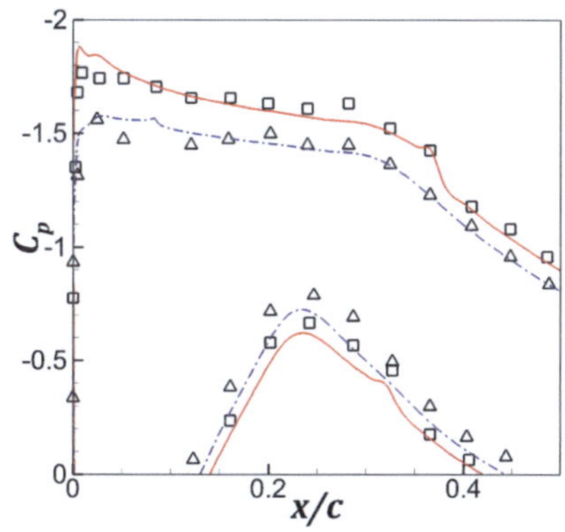

(b) Local Distribution

Fig. 4.20 Pressure coefficient distribution on the NREL-S814 airfoil at a 6° angle of attack

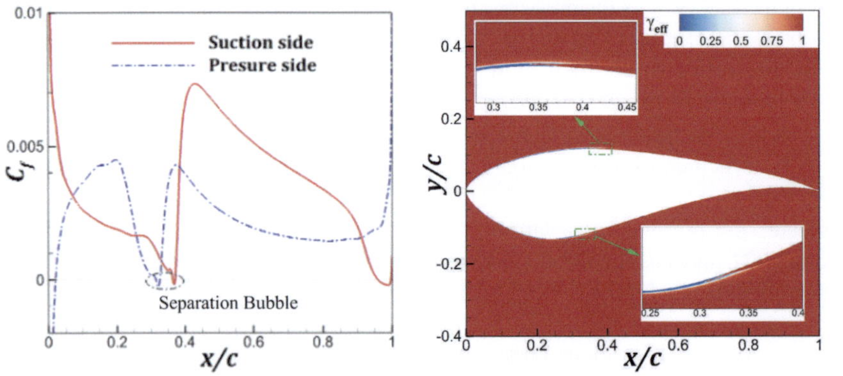

(a) Surface Friction Coefficient Distribution (b) Effective Intermittency Factor Distribution

Fig. 4.21 Distribution of surface friction coefficient and effective intermittency factor in the flow field for the smooth NREL-S814 airfoil at a 6° angle of attack

Fig. 4.22 Distribution of roughness amplification factor in the flow field for the rough NREL-S814 airfoil at a 6° angle of attack

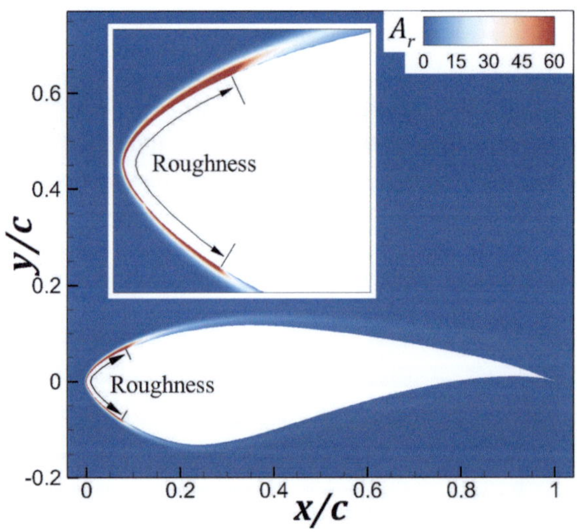

4.3.4 *Flow Around Roughened High-Pressure Turbine Blades*

At Karlsruhe Institute of Technology, Stripf et al. [219] carried out experimental studies in a heated wind tunnel to quantitatively assess the impact of surface roughness variations on external heat transfer and the laminar-to-turbulent transition for high-load turbine blades. The research encompassed an investigation of diverse roughness configurations in comparison to a smooth reference surface. This was conducted across a range of Reynolds numbers and inflow turbulence intensities. The experimental setup for the high-pressure turbine (HPT) blades is depicted in

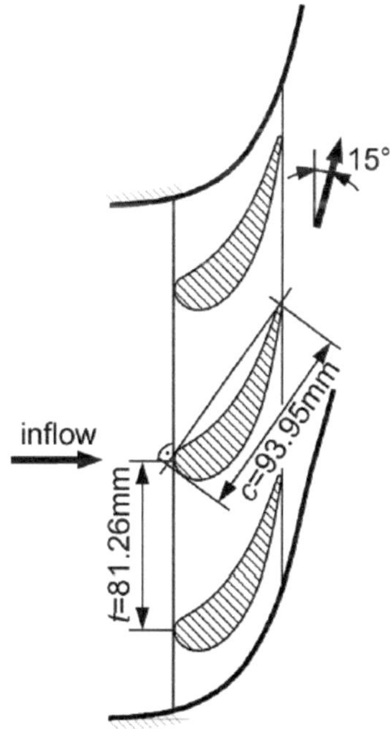

Fig. 4.23 Schematic diagram of high-pressure turbine blade cascade [220]

Fig. 4.23. The dimensions of the high-pressure turbine blade are as follows: a length of 93.95 mm, a chord length of 53 mm along the axis, and a blade pitch of 81.26 mm. The incoming flow angle of attack was set to 0°.

For the numerical simulation, a single blade was selected, with the contours of the computational domain and boundary conditions illustrated in Fig. 4.24. Periodic boundary conditions were applied to the upper and lower boundaries. Additionally, the figure illustrates the distribution of the mesh in the vicinity of the blade's leading and trailing edges. The computational domain comprised approximately 67,000 mesh elements, with 545 grid points distributed circumferentially around the blade. The y^+ value of the first-layer mesh was maintained below 1.0 to ensure accurate boundary layer resolution.

Stripf et al. [219–221] undertook extensive experimental investigations on high-pressure turbine blades, they explored various Reynolds numbers and turbulence intensities. This section selects two Reynolds numbers (1.4×10^5 and 2.5×10^5) and two turbulence intensities (3.5% and 8.0%) for detailed simulation. The specific computational conditions are summarized in Table 4.4. The Reynolds numbers are derived from the blade's actual chord length and the incoming flow velocity. In conducting heat transfer simulations, the blade wall temperature is maintained at 300 K, with the inlet air temperature set at 370 K. The Prandtl numbers for laminar and turbulent conditions are set to 0.72 and 0.86, respectively. In addition to simulations of smooth blades, four blades with varying roughness heights are subjected to analysis. The roughness parameters and corresponding heights are detailed in Table 4.5.

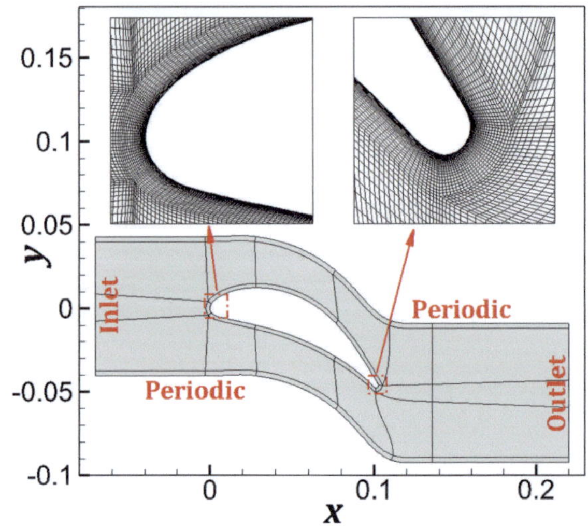

Fig. 4.24 Computational domain and local mesh of high-pressure turbine blade

Based on their experimental findings, Stripf et al. [106, 107] developed a roughness transition criterion and proposed a discrete element model tailored for random rough surfaces. Moreover, Ge and Durbin [108] used this roughness transition prediction model to assess transition phenomena on high-pressure turbine blades. Consequently, the results obtained from our transition model will be subjected to a critical comparison with the experimental data and findings derived from the aforementioned two established numerical methods.

As the experimental data on heat conduction across the blade surface is presented using the Nusselt number, it is essential to provide a brief explanation of this concept. The Nusselt number, named after the German physicist Wilhelm Nusselt, is a dimensionless metric that reflects the intensity of convective heat transfer relative to conductive heat transfer. The Nusselt number provides a means of measuring the ratio of convective to conductive heat transfer, thereby offering insight into the efficiency of

Table 4.4 High-pressure turbine calculation conditions

Case	Ma_{in}	Re_c	Tu_{in}	R_T
HPT C1	0.08	1.4×10^5	3.5	8.0
HPT C2	0.08	1.4×10^5	8.0	12.0
HPT C3	0.12	2.5×10^5	3.5	8.2
HPT C4	0.12	2.5×10^5	8.0	12.5

Table 4.5 Roughness parameters for high-pressure turbine blades

Roughness name	HPTV 20a	HPTV 30	HPTV 40b	HPTV 70
$k_s(\mu m)$	72	108	129	238

convective heat exchange processes.

$$Nu = \frac{\text{Convective heat transfer}}{\text{Conductive heat transfer}} = \frac{h \cdot L}{\kappa} \quad (4.20)$$

In the equation, L represents the characteristic length scale of the heat transfer surface, typically the dimension perpendicular to the surface, such as the diameter of a heat pipe or the thickness of a heat transfer layer. h denotes the convective heat transfer coefficient of the surface, while κ is the thermal conductivity of the stationary fluid.

In the experiments conducted by Stripf et al., the Nusselt number was determined using the blade chord length as the characteristic length for heat transfer. The calculation formula is as follows:

$$Nu_c = \frac{h \cdot c}{k} = \frac{q}{T_{tot} - T_w} \frac{c}{\kappa} \quad (4.21)$$

where T_{tot} is the total temperature of the inlet gas, T_w is the blade wall temperature, and q is the heat flux on the wall. For ease of comparison with experimental data, the Nusselt number predicted by the new model is also calculated using Eq. 4.21.

The distribution of the Nusselt number on the suction surface of the high-pressure turbine blade is shown in Fig. 4.29. The outcomes for varying roughness heights, under identical Reynolds numbers and incoming turbulence intensities, are represented on a single graph. As can be observed in the figure, the predicted transition location and the Nusselt number distribution align closely with the experimental data. Moreover, an increase in roughness height results in a downstream shift of the transition location on the suction surface, with the boundary approaching the leading edge. A comparison of Fig. 4.25a and b reveals that at relatively high roughness, the sensitivity of the transition onset point to the free-stream turbulence intensity is diminished. Similarly, at high incoming turbulence intensity, the sensitivity of the transition onset point to roughness height is reduced. Furthermore, a comparison of Fig. 4.25a and c or Fig. 4.25b and d reveals that the transition point on the rough surface is more susceptible to changes in the chord Reynolds number than on the smooth surface. The thermal conductivity calculated by Ge et al.'s model is highly sensitive to roughness height. As the roughness height increases, the Nusselt number in the turbulent region increases rapidly. In contrast, the results obtained from the model developed in this study are comparable to the experimental data and the model results presented by Stripf et al. This indicates that the predicted thermal conductivity is less sensitive to roughness height. In certain instances of roughness height and inflow conditions, the predicted rise in the Nusselt number curve along the length of the blade is shorter than that observed in the experimental data, indicating that the predicted transition develops at a faster rate than in the experiment.

Under low incoming turbulence intensity conditions, the new transition model detected a laminar separation bubble on the suction surface of a smooth blade. This occurrence was also observed in experiments, as depicted in Fig. 4.25. Figure 4.26 shows the distribution of the effective intermittency factor near the blade and the streamlines around the separation bubble. Despite the laminar separation bubble on

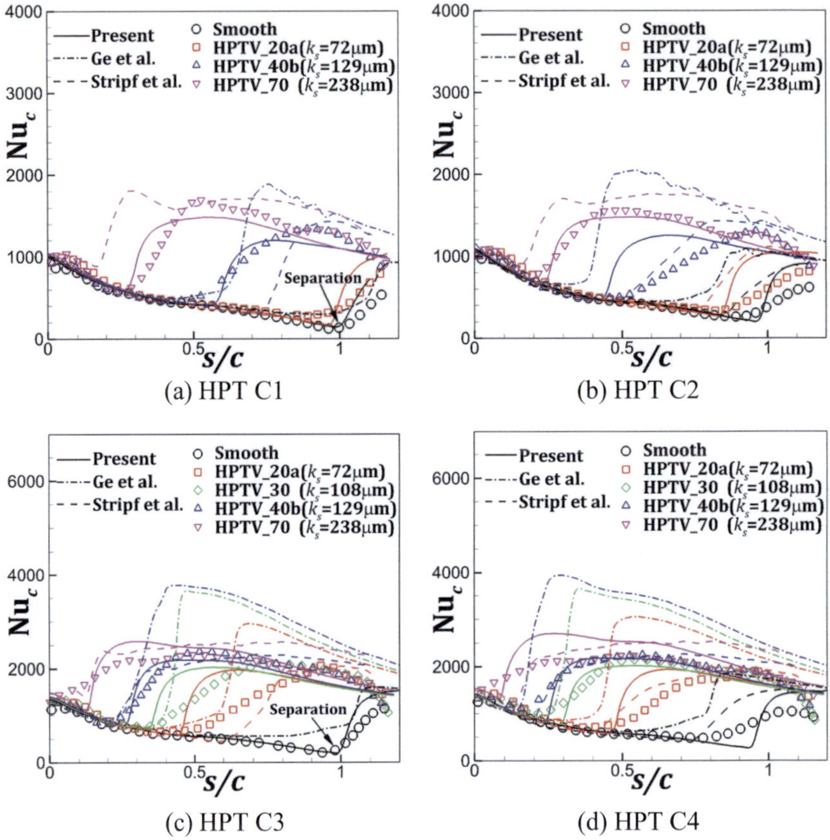

Fig. 4.25 Distribution of nusselt number on the suction surface of a high-pressure turbine blade

the blade's suction surface being relatively diminutive, its existence gives rise to a separation-induced transition. Conversely, bypass transition occurs on the suction surface of rough blades. The new model is thus demonstrated to be capable of accurately predicting both of these distinct transition mechanisms in turbomachinery flows.

Figures 4.27 and 4.28 illustrate the distributions of A_r, F_{onset}, γ_{eff} and turbulent kinetic energy around the blade in the HPT C2 case. To provide a more detailed comparison, the variable distributions near the transition region are shown in an enlarged view. The data demonstrate that, in the case of a smooth blade, the roughness amplification factor remains at zero throughout the entire flow field. An increase in the F_{onset1} function is observed in the vicinity of the trailing edge on the suction surface of the blade. As the source terms in the γ transport equation are triggered, the effective γ_{eff} value begins to rise, thereby initiating the flow transition. Consequently, the turbulence kinetic energy increases rapidly downstream of the transition point.

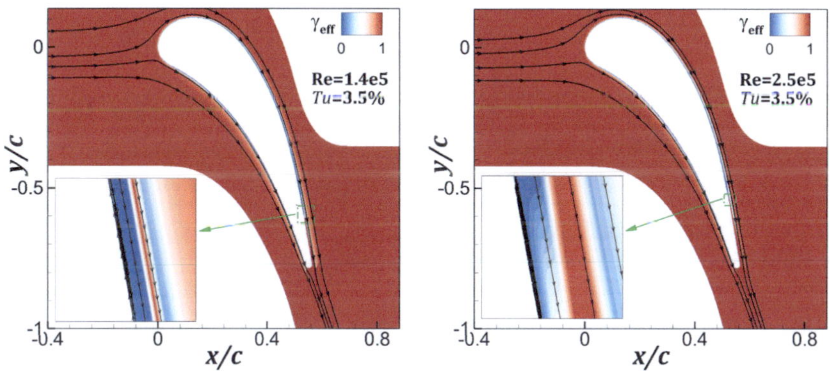

(a) HPT C1 (b) HPT C3

Fig. 4.26 Laminar separation bubble on the suction surface of a smooth high-pressure turbine blade

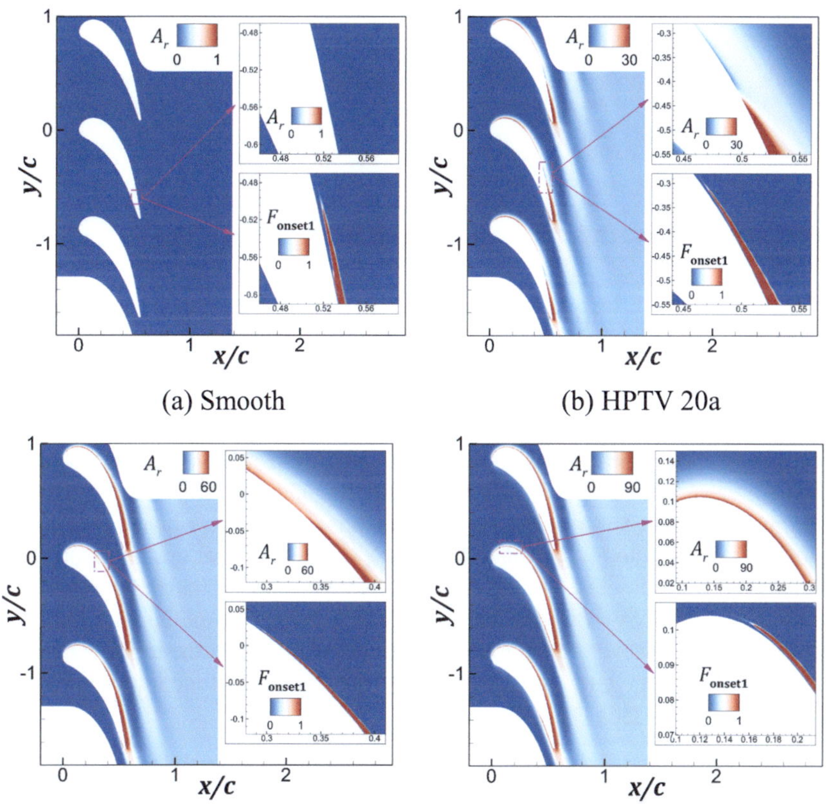

(a) Smooth (b) HPTV 20a

(c) HPTV 40b (d) HPTV 70

Fig. 4.27 Distribution of A_r and F_{onset1} in HPT C2 case

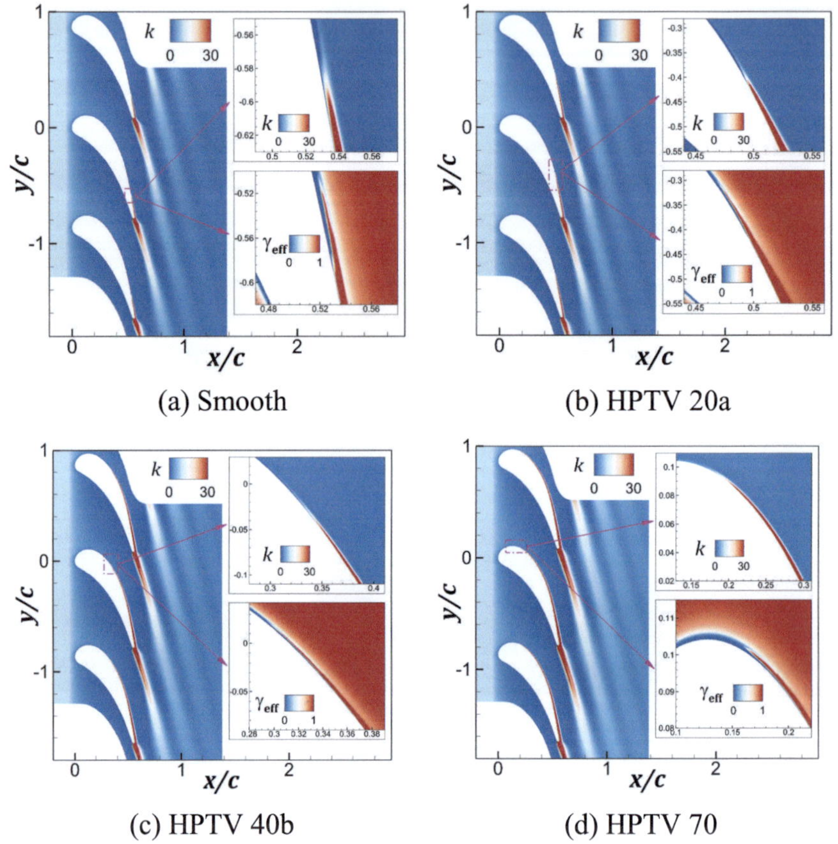

Fig. 4.28 Distribution of κ and γ_{eff} in HPT C2 case

For rough blades, the wall A_r value is greater than zero, and the transport equation causes A_r to propagate and diffuse within the boundary layer. As the roughness height increases, the A_r value rises within the laminar boundary layer. This causes the locations where the F_{onset1} function and γ_{eff} increase to shift upstream, bringing the transition location closer to the leading edge. For transitionally rough surfaces (HPTV 20a and HPTV 40b), the turbulent kinetic energy in the laminar region is essentially zero. However, for fully rough surfaces (HPTV 70), the turbulent kinetic energy is greater than zero, according to the wall correction function proposed by Knopp et al. (Fig. 4.28d).

4.3.5 Flow Around Rough Low-Pressure Turbine Blades

Stripf et al. [220, 221] carried out comprehensive experimental research on low-pressure turbine cascades to investigate how surface roughness affects the heat transfer characteristics of the blades. The experiments were conducted under different Reynolds numbers and varying levels of incoming turbulence intensity, with the roughness geometries on the blade surfaces classified into several categories. The setup for the experimental blades is depicted in Fig. 4.29, where the actual chord length of the blades is 113.34 mm, the axial chord length is 100 mm, and the blade pitch is 83.4 mm.

A blade is selected for numerical simulation, with the computational domain's contour and boundary conditions depicted in Fig. 4.30. The upper and lower boundaries are assigned periodic boundary conditions, while the left and right boundaries serve as the velocity inlet and pressure outlet, respectively. The figure also illustrates the mesh distribution near the leading and trailing edges. The computational domain comprises approximately 71,000 mesh elements, with 581 circumferential grid points surrounding the blade. The y^+ value for the first layer of the mesh is maintained below 0.3.

This section presents the results of numerical simulations and analyses on a low-pressure turbine blade with a single smooth surface and three distinct rough surfaces. The simulations were conducted at Reynolds numbers of 2.5×10^5 and

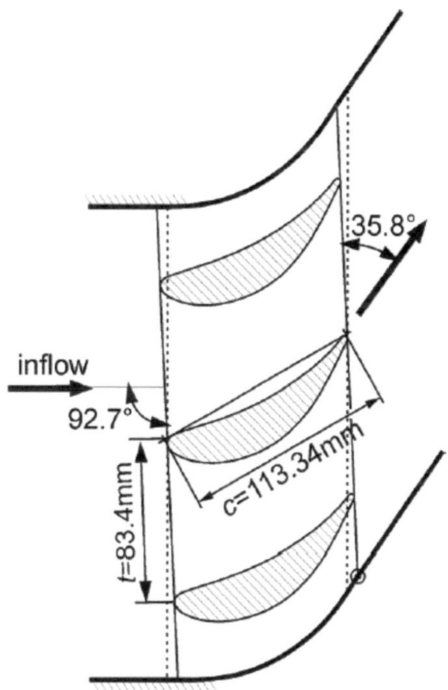

Fig. 4.29 Schematic diagram of the low-pressure turbine blade cascade [220]

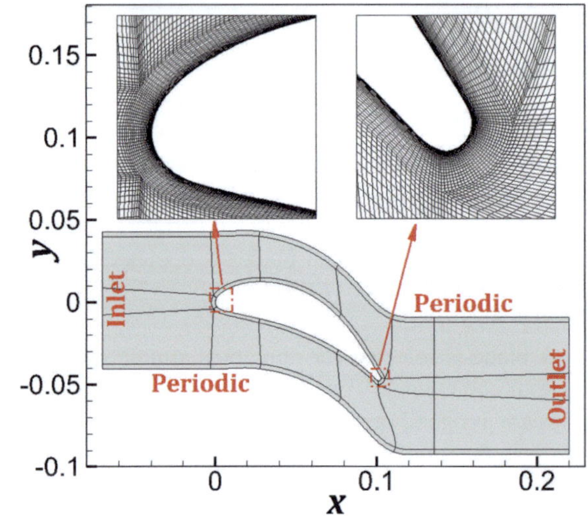

Fig. 4.30 Computational domain and local mesh of the low-pressure turbine blade

4.5×10^5, with incoming flow turbulence intensities of 4.6% and 10.0%, respectively (Table 4.6). The Reynolds numbers are derived from the actual chord length of the blade and the incoming flow velocity. The specific computational settings for the low-pressure turbine blade are outlined in a table within this section. Table 4.7 provides a detailed account of the roughness characteristics of the different blade surfaces.

The Nusselt number distribution on the low-pressure turbine blade is illustrated in Fig. 4.31. For comparison, the figure also includes experimental data alongside numerical results from Ge et al. and Stripf et al. The predictions from the present model accurately capture the effect of surface roughness on transition. In the case of the low-pressure turbine blade, an increase in roughness height results in a gradual upstream shift in the transition point. A comparison of Fig. 4.31a with Fig. 4.31c and Fig. 4.31b with Fig. 4.31d demonstrates that the Reynolds number exerts a considerable influence on the transition point for both smooth and rough blades. As

Table 4.6 Computational conditions for the low-pressure turbine

Case	Ma_{in}	Re_c	Tu_{in}	R_T
LPT C1	0.12	2.5×10^5	4.6	9.0
LPT C2	0.12	2.5×10^5	10.0	14.0
LPT C3	0.20	4.5×10^5	4.6	9.5
LPT C4	0.20	4.5×10^5	10.0	15.0

Table 4.7 Rough surfaces of the low-pressure turbine blade

Roughness name	LPTV 27a	LPTV 50a	LPTV 90
$k_s(\mu m)$	56	103	282

Reynolds number increases, the transition point tends to shift upstream. Besides, for blades with smooth surfaces or lower roughness heights, the initial transition point is particularly susceptible to fluctuations in the turbulence intensity of the incoming flow. Nevertheless, as the roughness height increases, the sensitivity of the transition point on the blade's suction side to incoming flow turbulence decreases. This trend is clearly demonstrated by a comparison of Fig. 4.31a with Fig. 4.31c and Fig. 4.31b with Fig. 4.31d. In the majority of cases, the onset of transition predicted by the model, which accounts for rough wall effects, is in good agreement with experimental data. In a few instances, the simulations demonstrate a delayed transition point and a shorter transition region in comparison to the experimental results. This discrepancy may be attributed to the model's insufficient consideration of the effects of roughness element density and streamline curvature.

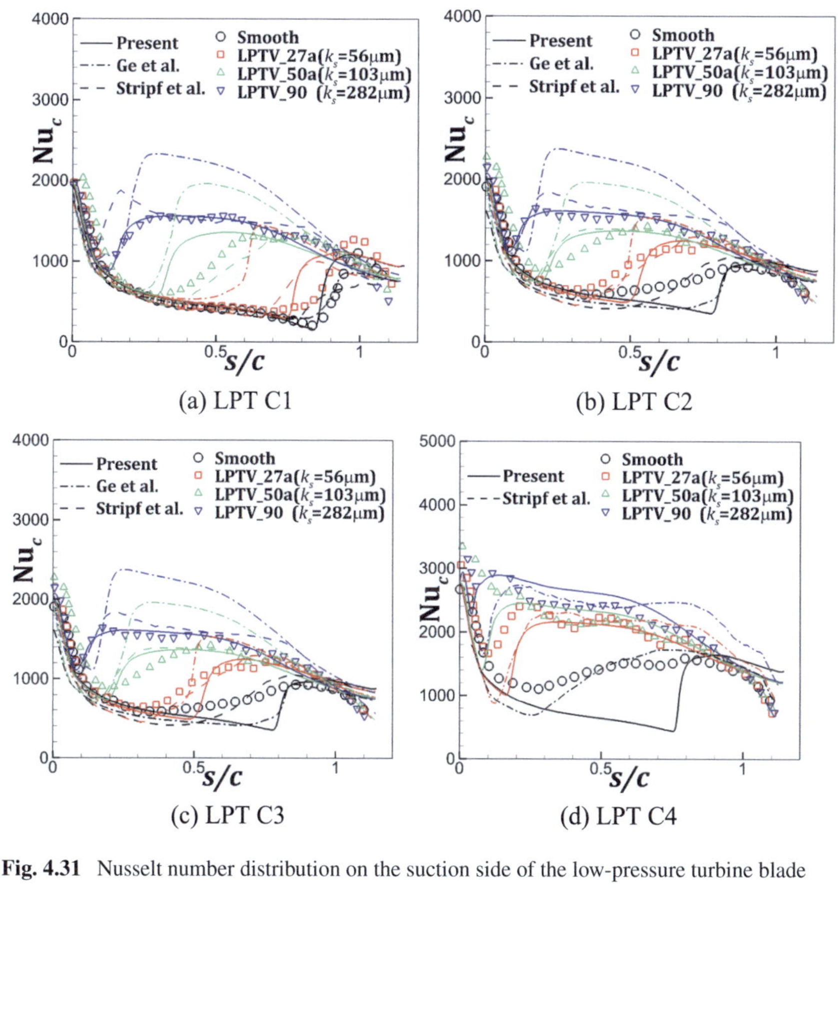

Fig. 4.31 Nusselt number distribution on the suction side of the low-pressure turbine blade

4.4 Summary

This chapter introduces a $\gamma - A_r$ transition model that incorporates the effects of rough walls. The model preserves the characteristics of the initial transition mode, with all variables solved locally. This simplifies the implementation of the mode in solvers and enhances scalability for large-scale parallel computations. The $\gamma - A_r$ model effectively captures the distribution of roughness amplification factors within the boundary layer and accurately reflects the impact of surface roughness on flow transition. As the height of the roughness increases, the onset of the transition shifts upstream. The model's predictions of transition locations on rough plates with different pressure gradients are in accordance with experimental data, thereby confirming the model's capacity to represent the influence of pressure gradients with precision. In flows through turbine blades, the model's predictions are in good agreement with experimental results for both transition induced by separation bubbles on smooth surfaces and bypass transition on rough surfaces. This indicates that the model has broad applicability in industrial flow scenarios. In conclusion, the $\gamma - A_r$ roughness transition model is a reliable predictor of transition processes on rough surfaces across a range of Reynolds numbers, pressure gradients, and turbulence intensities, making it suitable for simulating transition flows in engineering applications such as rough wings and turbomachinery. The model is primarily concerned with boundary layer flows over distributed rough surfaces, employing a statistical methodology to investigate the impact of roughness element disturbances. Nevertheless, the model does not take into account the specific effects of individual large discrete roughness elements on the flow.

Chapter 5
Modeling and Application of the VLES Method Considering the Influence of Transition and Surface Roughness

Abstract This chapter integrates the roughness-transition model with the VLES methodology, developing a hybrid prediction model for transition-separated flows applicable to rough walls. Validated through representative test cases including the V103 compressor cascade and Pak-B turbine blade, the proposed approach achieves simulation accuracy comparable to LES methods while demonstrating significantly reduced computational costs.

5.1 Introduction

In the field of engineering flow analysis, the RANS and URANS methods are the most commonly employed simulation tools. In the case of flow fields exhibiting substantial flow separation, the predictive precision of these methodologies is often inadequate, particularly with regard to the capture of unsteady phenomena. Hence, there has been a shift towards the adoption of hybrid RANS/LES methodologies, such as DDES, SAS, and PANS, with the aim of enhancing the reliability of flow field simulations. The transition from laminar to turbulent flow is a pervasive phenomenon in practical flow fields, with profound implications for aerodynamic performance. To illustrate, the transition process occurring within the boundary layer on the leeward side of an airfoil has the potential to exert a significant influence on its stall characteristics. Similarly, the occurrence of a separation bubble-induced transition on the pressure side of compressor blades can result in the onset of large-scale flow separation, which has the potential to significantly compromise the efficiency of the compressor. Moreover, variations in the transition point within the boundary layer can impact the extent of the wake separation zone surrounding aircraft landing gear and cylindrical struts, which can subsequently influence the level of aerodynamic noise generated. Although the aforementioned hybrid methods exhibit robust capabilities for simulating separated flows, they were predominantly developed under the assumption of fully turbulent conditions. Thus, they may not adequately capture the impact of boundary layer transition on the development of separated flows. In light of these deficiencies, research has increasingly concentrated on the creation of

© The Author(s), under exclusive license to Springer Nature Singapore Pte Ltd. 2025
J. Xu et al., *CFD-Compatible RANS/LES Modeling of Transitional and Separated Flows*, https://doi.org/10.1007/978-981-96-6886-1_5

hybrid RANS/LES methods that explicitly incorporate the effects of transition into their modeling frameworks.

In their study, Sørensen et al. [222] integrated the $\gamma - \overline{Re_{\theta t}}$ transition model with the DES method, as developed by Strelets [35] in the EllipSys3D software. Numerical simulations were conducted to investigate the flow around a cylinder across a Reynolds number range from 10 to 1×10^6, as well as the flow around the DU-96-W351 airfoil at high angles of attack. The "DES/transition" method showed better predictive capacity for the "drag crisis" in comparison to the pure DES method. This hybrid method accurately captured the relevant flow phenomena at the specified Reynolds numbers, including laminar separation, turbulent reattachment, and turbulent separation, and effectively tracked the variation of the separation point with Reynolds number. Nevertheless, Sørensen et al. did not present a comprehensive account of the interrelationship between the transition model and the DES method.

The $\gamma - \overline{Re_{\theta t}}$ transition model was incorporated into the Scale-Adaptive Simulation (SAS) model by You and Kwon [223], thereby enabling the new model to effectively capture both the laminar-to-turbulent transition in the boundary layer and the unsteady flow structures in the wake region. As both models are based on the $k - \omega$ SST framework, their governing equations exhibit notable similarities. The SAS model introduces an additional source term, Q_{SAS}, into the ω transport equation, while the transition model modifies the production and destruction terms in the k transport equation. This allows for the integration of these supplementary terms into a unified governing equation, provided that the activation of these terms occurs in distinct regions of the flow field and does not interfere with each other. The effective intermittency factor γ_{eff} undergoes rapid changes in proximity to the wall, yet it remains at 1.0 in the free stream. This suggests that the adjustments to the production and destruction terms in the k transport equation have a minimal impact on the flow away from the wall. Conversely, the additional source term Q_{SAS} is primarily activated in regions of unsteady wake. As a result, there is a discrepancy between the regions affected by each modification. In practical applications, the distinction between these regions is not always well-defined, necessitating the use of a blending function F_{blend} to manage and smoothly transition between the activation of the two sets of corrections. The equation forms proposed by You and Kwon are detailed as follows:

$$\frac{\partial}{\partial t}\begin{bmatrix} \rho k \\ \rho\omega \end{bmatrix} + \frac{\partial}{\partial x_j}\begin{bmatrix} \rho u_j k \\ \rho u_j \omega \end{bmatrix} - \frac{\partial}{\partial x_j}\begin{bmatrix} (\mu_l + \sigma_k \mu_t)\frac{\partial k}{\partial x_j} \\ (\mu_l + \sigma_\omega \mu_t)\frac{\partial \omega}{\partial x_j} \end{bmatrix} = \begin{bmatrix} \overline{P_k} - \overline{D_k} \\ P_\omega - D_\omega + CD_{SST} + Q_{SAS}(1 - F_{\text{blend}}) \end{bmatrix} \tag{5.1}$$

In simulations of flow around a cylinder using the SAS model, it was observed that the values of the functions F_1 and F_2 in the $k - \omega$ SST model remained excessively close to 1.0 over an extensive range in the wake region. This presents a challenge, as the wake flow is distinguished by pronounced unsteadiness and vortex shedding, which in turn necessitates the activation of supplementary source terms. It can be concluded that F_1 and F_2 are unsuitable as coupling functions for this model. In order to address the issue of excessive turbulence kinetic energy in the wake region, You and Kwon put forth the suggestion of introducing new coupling functions, which are defined as follows:

$$F_{\text{blend}} = \tanh\left(\left(\frac{C_{\text{blend}}\,\mu}{\rho\omega d_1^2}\right)^4\right),\ C_{\text{blend}} = 10,000 \tag{5.2}$$

In the aforementioned equations, the constant C_{blend} is calculated based on the boundary layer thickness of the cylinder, which was measured experimentally by Celik and Patel [224]. This value guarantees that the model accurately simulates the transition in the boundary layer of the cylinder while correctly activating the LES method in the wake region. Nevertheless, it should be noted that the aforementioned constant value of 10,000 may not be entirely suitable for non-cylindrical flows or for cylindrical flows under different flow conditions.

Qiao et al. [225] integrated the $\gamma - \overline{Re_{\theta t}}$ transition model into both DES and DDES frameworks. The turbulent kinetic energy transport equation in the modified model is expressed as follows:

$$\begin{aligned}\frac{\partial(\rho k)}{\partial t} + u_i\frac{\partial(\rho k)}{\partial x_i} &= \gamma_{\text{eff}}P_k - F_{\text{DES}}\cdot\min\left[1, \max\left(\gamma_{eff}, \gamma_{D_k}\min\right)\right]\cdot D_k \\ &+ \frac{\partial}{\partial x_i}\left[\left(\mu_l + \frac{\mu_t}{\sigma_k}\right)\frac{\partial k}{\partial x_i}\right].\end{aligned} \tag{5.3}$$

When F_{DES} is 1.0 in the near wall region, the model operates as a transition model. Nonetheless, in the separated flow regions away from the wall, where the flow has become fully turbulent, γ_{eff} reaches 1.0, thus entering the DES or DDES model.

Sa et al. [226] developed a model-switching function, designated as f_{Tran}, based on $F_{\theta t}$ and coupled the $\gamma - \overline{Re_{\theta t}}$ transition model with the DDES model. Subsequent to the introduction of the coupling function, the destruction term in the turbulent kinetic energy transport equation of the novel model is reformulated as follows:

$$\tilde{D}_{k,\ \text{DDES-Tran}} = \frac{\rho k^{\frac{3}{2}}}{\tilde{l}_{\text{Tran}}} \tag{5.4}$$

$$\tilde{l}_{\text{Tran}} = l_{k-\omega} - f_d f_{\text{Tran}}\max(0, l_{k-\omega} - C_{DES}\Delta) \tag{5.5}$$

$$f_{\text{Tran}} = \left(1 - F_{\theta t}^{\text{DDES-Tran}}\right) \tag{5.6}$$

$$F_{\theta t}^{\text{DDES-Tran}} = \left(\frac{(\min(\max(F_{\theta t}, \chi), 1.0) - \chi)}{(1-\chi)}\right)^2 \tag{5.7}$$

In this equation, f_d is equal to 1 minus F_{SST}, where F_{SST} is selected from the mixing functions F_1 and F_2 of the SST turbulence model. The constant $\chi = 0.8$ is employed for the purpose of regulating the activation range of the RANS model in the wake region.

Hodara et al. [230] developed a new turbulence closure model within the additive hybrid filter framework for hybrid RANS/LES, as proposed by Germano [227, 228] and Sanchez-Rocha and Menon [229]. The integration of the localized dynamic kinetic energy model (LDKM) with the $\gamma - \overline{Re_{\theta t}}$ transition model allows for the precise capture of flow characteristics in large separated boundary layers within transitional flows. In the new model, hybrid turbulent kinetic energy is defined as a combination of turbulent kinetic energy from the URANS model and subgrid-scale kinetic energy from the LES mode.

$$\mathcal{K} = \mathcal{F}k + (1 - \mathcal{F})k^{sgs} \tag{5.8}$$

The transport equation for the hybrid turbulent kinetic energy $\mathcal{K}$ is given by:

$$\frac{\partial \rho \mathcal{K}}{\partial t} + \frac{\partial}{\partial x_j}\left(\rho u_j \mathcal{K}\right) = \frac{\partial}{\partial x_j}\left[\mu + \mathcal{F}\sigma_k \mu_T + (1 - \mathcal{F})\frac{\mu_{\text{sgs}}}{C_t}\right]\frac{\partial \mathcal{K}}{\partial x_j} + K^S \tag{5.9}$$

where

$$\begin{aligned} K^S = {} & \mathcal{F}\left[\gamma_{\text{eff}}\min\left(\tau_{ij}^{\text{rs}}\frac{\partial u_i}{\partial x_j},\ 10\beta^* \rho\omega\mathcal{K}\right) - \overline{\gamma}_{\text{eff}}\beta^* \rho\omega\mathcal{K}\right] \\ & + (1 - \mathcal{F})\left[\tau_{ij}^{\text{sgs}}\frac{\partial u_i}{\partial x_j} - \frac{C_{\epsilon}\rho K^{\frac{3}{2}}}{\Delta}\right] \end{aligned} \tag{5.10}$$

In Eq. 5.8, the mixing function $\mathcal{F}$ governs the transition from the URANS model to the LES mode. Hodara et al. defined it as follows:

$$\mathcal{F} = \max(F_2; F_4; F_5) \tag{5.11}$$

$$F_4 = 1 - \min\left[\max\left\{\exp\left[-\left(\frac{y}{2.0\delta}\right)^4\right], 1.0 - \left(\frac{\gamma - \frac{1}{c_{e2}}}{1 - \frac{1}{c_{e2}}}\right)^2\right\}, 1.0\right] \tag{5.12}$$

$$F_5 = 1 - \tanh(1.2\gamma)^8 \tag{5.13}$$

In this equation, F_2 represents the mixing function derived from the SST turbulence model, F_4 ensures that the URANS model effectively captures the decay of turbulence intensity in the free-stream flow, and F_5 ensures that the URANS model is maintained in laminar regions, preventing the inadvertent activation of the LES mode.

Coder and Ortiz-Melendez [231] devised a DDES method specifically designed for forecasting transitional flows. This involved integrating the SA turbulence model with the Amplification Factor Transport (AFT2017b) transition model [172]. The method was subsequently applied to simulate the MD30P3N three-element wing, and the results showed a high degree of concordance with the corresponding experimental data. A pivotal aspect of their model is the calibration of the DDES length scale, wherein an intermittency factor is incorporated to restrict the applicability of the DDES method in laminar boundary layers. The new length scale is defined as follows:

$$\tilde{d} = d - \gamma_{\text{lim}} f_d \max(0, d - C_{\text{DES}}\Delta) \tag{5.14}$$

where

$$\gamma_{\text{lim}} = \min\left[\exp(\tilde{\gamma}), 1\right] \tag{5.15}$$

In the equation, the modified intermittency factor $\tilde{\gamma}$ is the natural logarithm of the actual intermittency factor, which is obtained by solving its transport equation. This formulation guarantees that the model employs solely the RANS length scale within the laminar boundary layer.

Xiao et al. [232] developed an advanced DDES model based on the three-equation $k-\omega-\gamma$ transition model [233]. In this model, the transition process occurring within the attached boundary layer is captured by the transition model, while large-scale separation flows are addressed using a hybrid method. The concurrent operation of the transition model and the DDES model carries the potential for inaccurate predictions of the characteristic length scale to affect both laminar and transition regions. To mitigate this issue, Xiao et al. introduced a restriction mechanism to ensure that the transition model is applied exclusively to laminar and transition zones, while the DDES model is confined to fully turbulent regions. The characteristic length scale is adjusted using the intermittency factor γ, with the correction formula provided below:

$$l_{\text{DDES}} = l_{\text{RANS}} - \frac{1+\text{sign}(1.0, \gamma - 0.99)}{2.0} \times f_d \times \max\{0, (l_{\text{RANS}} - l_{\text{LES}})\} \tag{5.16}$$

Subsequently, the term denoting the destruction of turbulent kinetic energy is redefined in accordance with the adjusted length scale.

$$\varepsilon_k = \rho \frac{k^{\frac{3}{2}}}{l_{\text{DDES}}} \tag{5.17}$$

Kim and Kwon [234] coupled the $\gamma - \overline{Re_{\theta t}} - CF^{+}$ cross-flow transition model with the IDDES method. The equations for their coupled model are as follows:

$$\frac{D}{Dt}\begin{bmatrix} \rho k \\ \rho\omega \end{bmatrix} = \nabla\begin{bmatrix} (\mu_l + \sigma_k \mu_t)\nabla k \\ (\mu_l + \sigma_\omega \mu_t)\nabla\omega \end{bmatrix}$$

$$+\begin{bmatrix} (P_k - \beta * \rho k\omega \cdot F_{\mathrm{IDDES}}) \cdot \left(1 - \mathrm{trunc}\left(\tilde{f}_d\right)\right) \\ +(\gamma \cdot P_k - \beta^* \rho k\omega \cdot (\min(\max(\gamma, 0.1), 1.0))) \cdot \mathrm{trunc}\left(\tilde{f}_d\right) \\ \frac{\alpha}{v_t} P_k - \beta\rho\omega^2 + 2(1 - F_1)\frac{\rho\sigma_{\omega 2}}{\omega}\nabla k \cdot \nabla\omega \end{bmatrix} \tag{5.18}$$

The mixed approaches of "RANS transition models + LES modes", which were discussed in the preceding section, focus exclusively on the effects of smooth walls on flow transition and separation. Nonetheless, as previously discussed, surface roughness has a considerable impact on both transition and separation. Some studies have employed LES or DNS methods to explore the effects of isolated and distributed roughness elements on transition mechanisms [235–238], but these investigations are generally limited to simple geometries and specific regions. Other research has sought to develop empirical relationships for roughness-induced transition within the RANS framework by analyzing both experimental and numerical data [106–109]. However, such methods have often proven inadequate for predicting large-scale separation flows. To address this issue, a novel mixed RANS/LES method has been developed within the VLES framework, based on the $\gamma - A_r$ roughness transition model previously discussed. The objective of this new method is designed to accurately capture the impact of rough walls on both bypass transition and laminar separation bubble-induced transition, thereby providing a more effective and precise simulation of these phenomena.

5.2 Model Construction

In the construction of the $\gamma - A_r -$ VLES model, the form of the scale control function, which exerts a significant influence on the selection of the flow field model, is of paramount importance. The initial scale control function was devised in accordance with the f_{VLES} turbulence model. When VLES is coupled with the transition model, it is necessary to modify the aforementioned function in order to create a new scale control function, designated as f_{hybrid}.

5.2.1 *Laminar Region Model Limitation*

In the hybrid RANS/LES method, the LES mode is unable to simulate the transition process of the flow in the absence of artificially introduced disturbances. Hence, in this mode, the transition process of the flow is predicted using a RANS transition model, while the LES mode is activated in regions exhibiting significant separation. One of the most significant limitations of this model is the necessity to maintain the RANS model within the laminar region, without transitioning to the LES mode. This guarantees that the transition of the flow is initiated by the transition criteria that were previously established, thus enabling the model to accurately describe the transition process from laminar to turbulent flow. In accordance with this stipulation, it is imperative to establish an appropriate coupling function. This function should be equal to 1 within the laminar boundary layer and gradually decrease to 0 at the boundary layer edge and in the turbulent region. In the hybrid RANS/LES method, the blending functions F_1 and F_2, which are typically employed in the $k-\omega$ SST model, are frequently selected as the coupling functions between the RANS and LES modes. Nevertheless, in the laminar boundary layer, the functions F_1 and F_2 exhibit singular values less than 1 in certain regions of the boundary layer. Langtry and Menter [67] also addressed this issue in the transition model they developed and proposed an F_3 function to correct the function values within the laminar boundary layer. Figure 5.1 illustrates the flow field results of the T3A flat plate case, which were calculated using the $\gamma - A_r$ transition model. Each function was computed from the flow field data. For purpose of comparison, the flow field variables at $x/L = 0.3$ within the laminar region are extracted and presented in Fig. 5.2. It is obvious that the functions F_1 and F_2 exhibit values less than 1 in certain regions of the boundary layer. The F_3 function ensures that the $k-\omega$ model in the SST model is correctly activated in both the laminar and transition regions of the boundary layer. Accordingly, by invoking the F_3 function and the methods of Hodara et al. [230] and Xiao et al. [232], the RANS model within the laminar boundary layer is safeguarded to guarantee its accurate activation.

The coupling function for the purpose of safeguarding the laminar region is defined as follows:

$$F_{\text{lam}} = \frac{1.0-\text{sign}(\gamma-C_\gamma)}{2.0} \tag{5.19}$$

The intermittency factor γ is employed to differentiate between the regions where the transition model is operational and those where the VLES method is applied. The constant C_γ is set to 0.99.

The scale control function for the rough transition VLES model is defined as follows:

$$f_{\text{hybrid}} = \max(f_{\text{VLES}}, F_{\text{lam}}). \tag{5.20}$$

In the event that γ is less than 0.99, F_{lam} is set to 1, and the flow field is resolved using the RANS model. As a result, when γ exceeds 0.99, F_{lam} is set to 0, and the flow field is transitioned to the VLES model for resolution.

5.2.2 *Protection of Free-Stream Turbulence*

In the VLES method, the boundary layer protection function f_b undergoes a transition from a value of 1 in the vicinity of the wall (which indicates the RANS model) to a value of 0 in regions situated at a greater distance from the wall and in separated

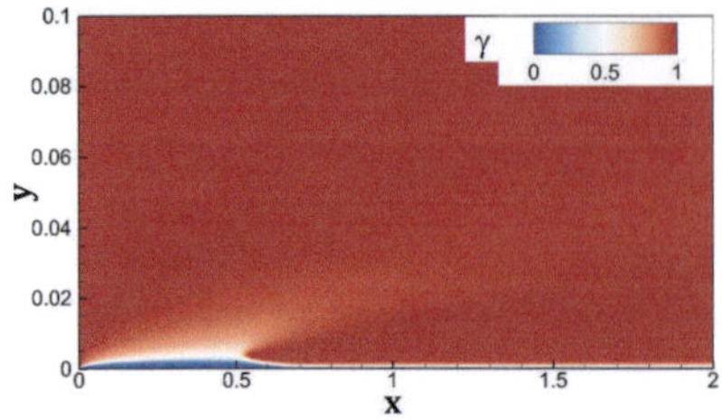

(a) Distribution of Intermittency Factor

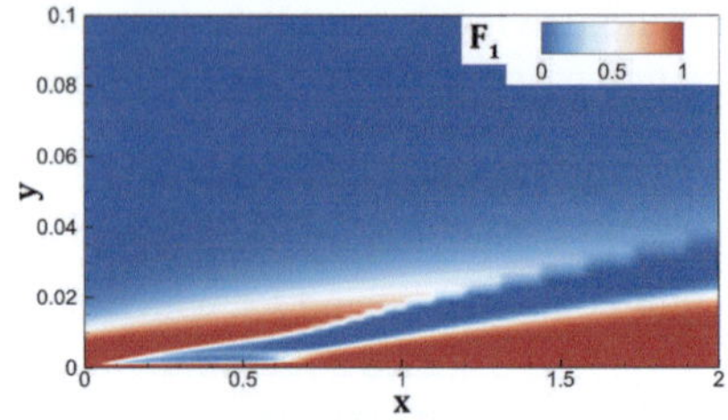

(b) Distribution of F_1 Function

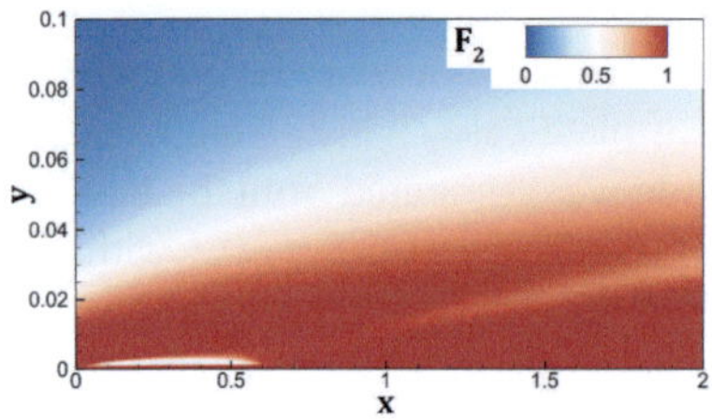

(c) Distribution of F_2 Function

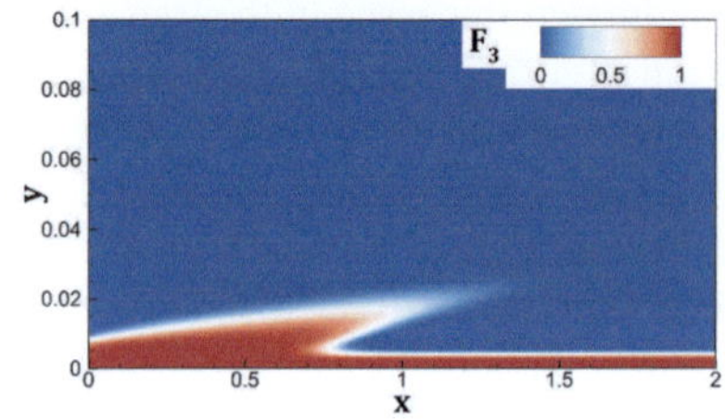

(d) Distribution of F_3 Function

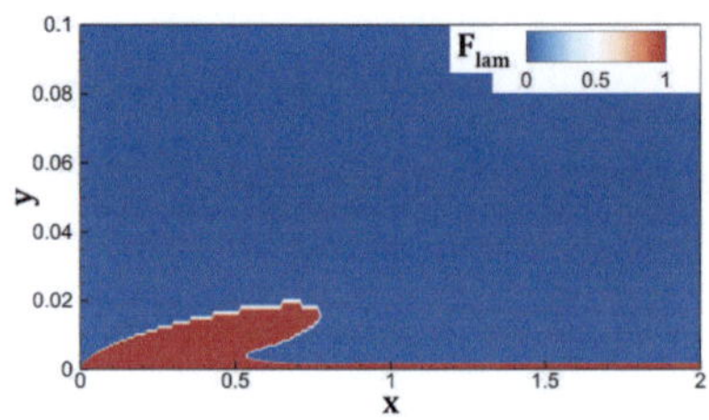

(e) Distribution of F_{lam} Function

Fig. 5.1 Distribution of intermittency factor and mixing function in T3A flat plate boundary layer

areas (which indicates the VLES model). However, there is a risk of inadvertently activating the VLES model in the free-stream region, which may lead to inaccuracies in the turbulence decay process if the grid resolution is insufficient or if the boundary conditions are not properly defined. In fully turbulent hybrid RANS/LES methods, this issue typically has a minimal impact on simulation results due to the lower sensitivity of turbulence models to boundary layer conditions. Nevertheless, in transition models, high turbulence intensity in the free stream can directly generate turbulent spots within the boundary layer, which may result in premature transition. It is therefore essential to accurately capture the decay of turbulence in order to make precise predictions regarding bypass transition. As the RANS transition model has been calibrated specifically for these types of flows, this study modifies the scale control function to ensure that the URANS model is used instead of the VLES model to accurately capture the decay of free-stream turbulence intensity. This necessitates the introduction of a free-stream indicator, which serves to identify the regions where the coupling function should revert to the RANS model (where f_{hybrid} approaches 1).

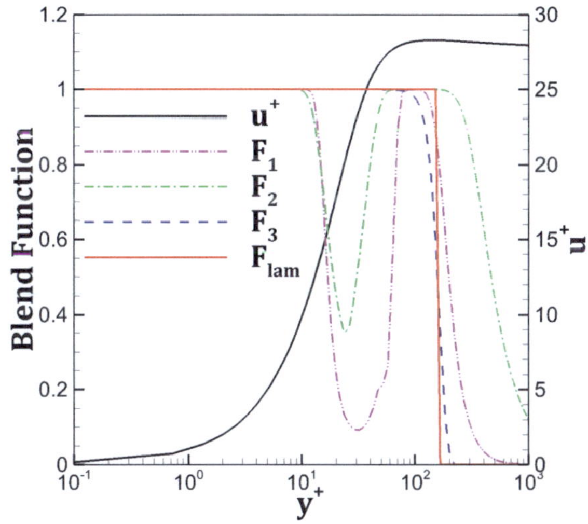

Fig. 5.2 Distribution of the mixing function at $x/L = 0.3$ in the boundary layer

The $F_{\theta t}$ function, initially proposed by Langtry and Menter [68], was selected and modified to ensure its continued efficacy in the wake regions of the flow. The new function is defined as follows:

$$F_{\text{free}} = 1 - \min\left[\max\left\{e^{-\left(\frac{y}{c_1 \delta}\right)^4}, 1.0 - \left(\frac{\gamma - \frac{1}{c_{e2}}}{1 - \frac{1}{c_{e2}}}\right)^2\right\}, 1.0\right] \tag{5.21}$$

In this context, c_1 is assigned a value of 2.0, while c_{e2} is set to 50.0.

The scale control function for the roughness-induced transition VLES model is rewritten as follows:

$$f_{\text{hybrid}} = \max(f_{\text{VLES}}, F_{\text{lam}}, F_{\text{free}}) \tag{5.22}$$

In the laminar and turbulent boundary layers, as well as in the wake region, the function F_{free} is set to zero. In contrast, it assumes a value of one in the far field and other areas. This configuration ensures that the scale control function f_{hybrid} is equal to 1 in the far field, thus maintaining the simulation in RANS model. Conversely, in the boundary layers and wake region, the simulation reverts to the VLES model.

5.2.3 Laminar Separation Bubble-Induced Transition Correction

In the transitional phase resulting from laminar separation bubbles, the flow initially maintains a laminar state at the point of separation. As the flow progresses, it undergoes a transition from laminar to turbulent behavior. Ultimately, the turbulent boundary layer reattaches and gives rise to the formation of the separation bubble. The $\gamma - A_r$ transition model utilizes an empirical correction function to account for the impact of turbulence intensity, pressure gradients, and surface roughness on transition caused by laminar separation bubbles. This correction function was developed and calibrated within the RANS framework. When a laminar separation bubble forms in the boundary layer, the correction intermittency factor γ_{sep} increases the local effective intermittency factor beyond 1.0. As previously stated, the novel VLES model uses the intermittency factor γ to differentiate between laminar and turbulent boundary layers within the coupling function. When the value of γ exceeds 0.99, the simulation is switched over to the VLES model. Consequently, the VLES model is activated at the onset of separation, which results in the RANS-based correction method for laminar separation bubble transition becoming less effective. This consequently results in the model's reduced capacity to accurately capture the effects of turbulence intensity and surface roughness, as well as hindering its ability to predict the transition process within the laminar separation bubble. As illustrated in Fig. 5.3, the VLES model is initiated when a laminar separation bubble emerges, although the transition process itself commences downstream. To address this issue, it is essential to maintain the RANS model during the initial stages of laminar separation boundary layer development. It is recommended that the simulation be switched to the VLES model only after the fluctuation intensity within the separation boundary

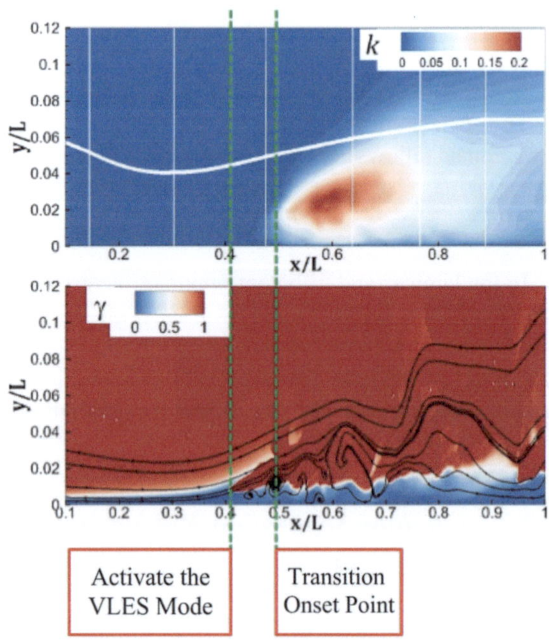

Fig. 5.3 Distribution of turbulent kinetic energy and intermittency factor in the flow field near the laminar separation bubble on the flat plate

layer reaches a certain threshold, in order to ensure an accurate representation of the transition process.

A new coupling function F_{sep} is introduced, defined as follows:

$$F_{\text{sep}} = F_{gs} F_{rt} \tag{5.23}$$

$$F_{gs} = \frac{1 + \text{sign}\left(1.0, \gamma_{\text{eff}} - C_{gs}\right)}{2.0} \tag{5.24}$$

$$F_{rt} = \mathrm{e}^{-(C_{rt1} R_T)^{C_{rt2}}} \tag{5.25}$$

In the equation, the viscosity ratio is defined as $R_T = \nu_t / \nu$. The model constants C_{gs}, C_{rt1}, and C_{rt2} have been set to 1.0, 1.1, and 10.0, respectively. When γ_{eff} is less than 1.0, the function F_{gs} is set to 0, preventing the activation of the coupling function F_{sep}. When γ_{eff} exceeds 1.0, F_{gs} becomes 1, activating the coupling function F_{sep} and ensuring that the RANS model is applied during the early stages of laminar separation bubble development. As the intensity of turbulent fluctuations increases, the function F_{rt} deactivates the coupling function F_{sep}, thereby transitioning the simulation to the VLES model.

The aforementioned model correction approaches, when combined, yield the following expression for the scale control function for the $\gamma - A_r -$ VLES hybrid method, which considers the effects of transition and surface roughness:

$$f_{\text{hybrid}} = \max\left(f_{\text{VLES}}, F_{\text{lam}}, F_{\text{free}}, F_{\text{sep}}\right). \tag{5.26}$$

It should be noted, however, that the aforementioned blending function is not without certain limitations. In particular, in the fully turbulent region following transition, the blending function continues to indicate the presence of an extensive laminar region, which is inconsistent with physical expectations. Hence, a new blending function is proposed in this work [239] to ensure an accurate representation of the fully turbulent region.

$$f_{\text{hybrid}} = \max\left(f_{\text{VLES}}, F_{\text{lam - new}}, F_{\text{free}}, F_{\text{sep}}\right) \tag{5.27}$$

$$F_{\text{lam - new}} = \min(F_{\text{lam}}, f_{\text{lam - mod}}) \tag{5.28}$$

$$f_{\text{lam - mod}} = \tanh\left(3.0\mathrm{e}^{-[0.2\text{Tu}(1-F_{\text{free}})]^4}\right) \tag{5.29}$$

As illustrated in the accompanying figure, the distribution of the blending function in the compressor cascade flow reveals that, after the improvement, the indication of laminar regions within the fully turbulent area is markedly compressed to the viscous sublayer. This adjustment is both physically accurate and reasonable (Fig. 5.4).

Fig. 5.4 Distribution of the blending function in compressor cascade flow

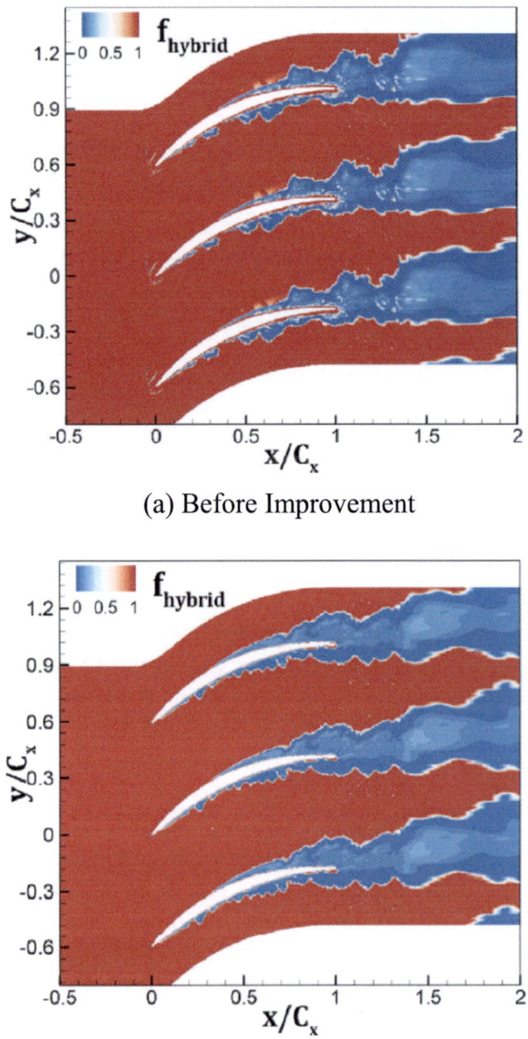

(a) Before Improvement

(b) After Improvement

5.3 Case Validation

In order to validate the predictive performance and simulation accuracy of the new hybrid method for flows affected by transition and surface roughness, the $\gamma - A_r -$VLES method was employed in numerical simulations of several test cases. These include a smooth flat plate with attached flow, a smooth flat plate with pressure-gradient-induced separation, a rough cylinder, a rough V103 compressor blade and a smooth Pak-B turbine blade.

5.3.1 Smooth Flat Plate Attached Flow

The recently developed VLES method, which is based on the $\gamma - A_r$ transition model, represents a significant advancement in the field of roughness-induced transitions. In the context of transitional flows in attached boundary layers on smooth flat plates, it is anticipated that the hybrid model will demonstrate consistent transition prediction capabilities with those of the original RANS transition model. To verify this, the $\gamma - A_r-$ VLES method was applied to simulate the T3 series flat plate cases, with a view to assessing the model's ability to predict the transition process accurately. The specific computational conditions for each smooth flat plate, including the inlet velocity, turbulence intensity, viscosity ratio, and molecular viscosity, are detailed in Table 5.1. The two-dimensional grid presented in Chap. 3 was extended to three dimensions, with a spanwise length of $0.12L$ and 37 grid points distributed along the span. Periodic boundary conditions were applied on both spanwise sides, while all other boundary conditions were maintained in accordance with the specifications outlined in Chap. 3.

In accordance with the inlet boundary conditions delineated in Table 5.1, the distribution of free-stream turbulence intensity and velocity along the streamwise direction for the flat plate flow field was calculated. Figure 5.5 depicts the variation of free-stream turbulence intensity with respect to streamwise position for the different flat plate cases. Additionally, Fig. 5.5b presents the variation of free-stream velocity along the streamwise direction for the T3C series flat plates. The outcomes yielded by the $\gamma - A_r-$ VLES method exhibit a high degree of concordance with the empirical data.

Figure 5.6 illustrates the surface friction coefficient distribution predicted by the $\gamma - A_r-$ VLES method. This distribution is then compared with both experimental data and the results obtained from the $\gamma - A_r$ transition model employed in this study. In all five cases involving flat plates with attached flow, the $\gamma - A_r-$ VLES method employs the RANS model for both laminar and turbulent boundary layers. The resulting surface friction coefficient distribution is in close agreement with that of the RANS transition model and aligns well with experimental measurements. It demonstrates that the $\gamma - A_r-$ VLES hybrid method effectively simulates the transition process in attached boundary layers, in agreement with the original RANS transition model.

Table 5.1 T3 series flat plate calculation conditions

Case	U_{inlet} (m/s)	Tu_{inlet} (%)	R_T	$\nu(\text{m}^2/\text{s}^{-1})$
T3A	5.4	3.3	12.0	1.5×10^{-5}
T3B	9.2	6.2	100.0	1.5×10^{-5}
T3C2	5.5	2.8	10.0	1.5×10^{-5}
T3C3	4.1	3.0	6.0	1.5×10^{-5}
T3C5	9.2	3.5	17.0	1.5×10^{-5}

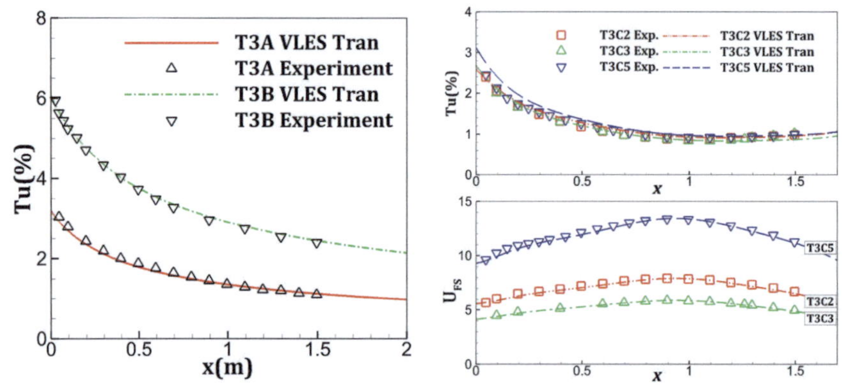

(a) T3A and T3B Flat Plates (b) T3C Flat Plate

Fig. 5.5 Variation of free-stream turbulence intensity and velocity along the streamwise position for the T3 series flat plates

The comparative results indicate that the recently constructed scale control functions, designated as F_{lam} and F_{free}, are effectively fulfilling their intended functions. The function F_{lam} serves to safeguard the laminar boundary layer, whereas F_{free} is responsible for ensuring the precision of free-stream turbulence intensity.

5.3.2 Smooth Flat Plate Separation Flow

Subsequently, the efficacy of the novel hybrid method in forecasting the transition of attached flow flat plates was evaluated. To this end, the performance of the method in simulating separated flow transitions was validated using the pressure gradient-separated flat plate case discussed in Chap. 3.

In the L&H flat plate case from Chap. 3, when the turbulence intensity prior to the formation of separation bubble is 1%, the resulting separation bubble is relatively large. A discrepancy is observed between the flow separation patterns predicted by the RANS transition model and the experimental results and LES data. In this section, the advanced $\gamma - A_r -$ VLES method is employed to conduct a numerical simulation of this case.

Lou and Hourmouziadi [178] conducted experimental investigations on this flat plate flow in the low-speed wind tunnel at the Berlin Institute of Technology. Wissink and Rodi [44], Lardeau et al. [46], and Yin and Durbin [239] employed high-fidelity numerical methods to simulate this transitional separated flow. In this study, we utilise data from Lardeau et al. who employed the dynamic Smagorinsky LES method [46], and compare it with predictions from the $\gamma - A_r$ transition model and the $\gamma - A_r -$ VLES method. The configuration of the flat plate model and the computational

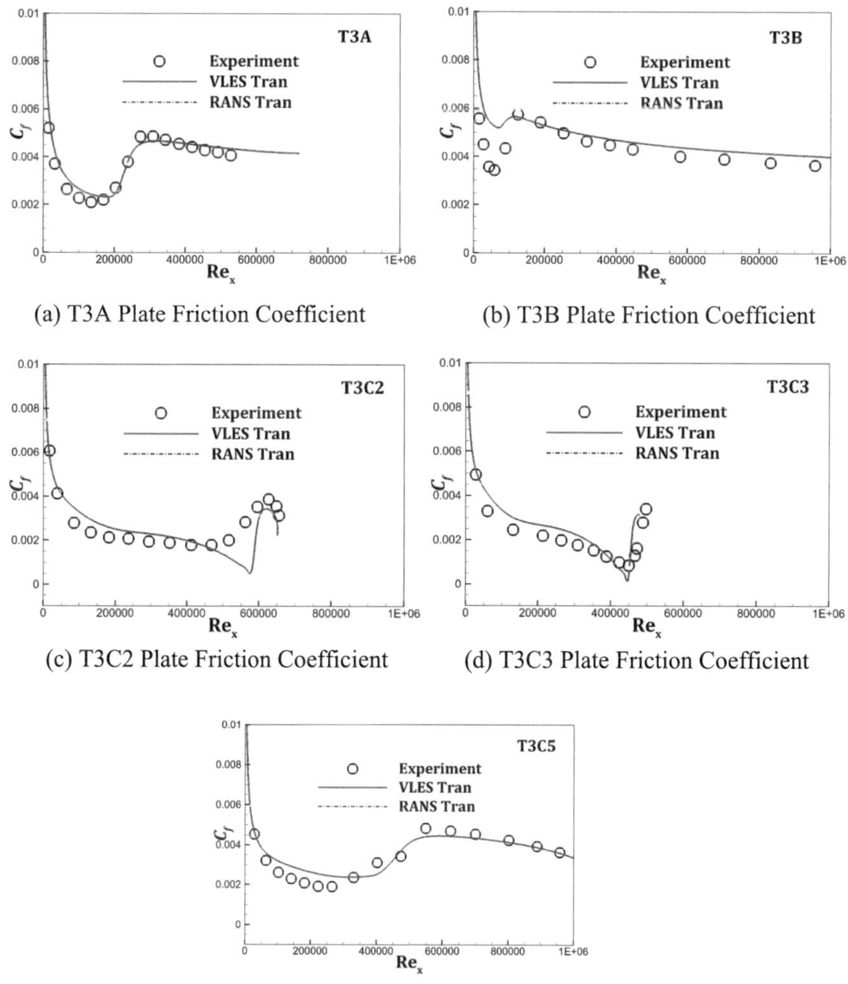

(a) T3A Plate Friction Coefficient

(b) T3B Plate Friction Coefficient

(c) T3C2 Plate Friction Coefficient

(d) T3C3 Plate Friction Coefficient

(e) T3C5 Plate Friction Coefficient

Fig. 5.6 Distribution of friction coefficients for the T3 series plates

domain are illustrated in Figs. 5.7 and 5.8, respectively. In order to create an appropriate pressure gradient distribution along the streamwise direction, a curved slip surface is used on the upper wall, as proposed by Wissink and Rodi [44]. The computational domain is extended to a length of 0.5 times the plate length (L) upstream of the plate and 1.5 times the plate length ($x/L = 1.5$) downstream, allowing the flow to fully reattach and develop into a fully turbulent state. The spanwise length of the computational domain is 0.12 times the length of the plate (L). The grid resolution in the three directions is set to $(n_x, n_y, n_z) = (481, 113, 61)$.

Fig. 5.7 L&H plate model

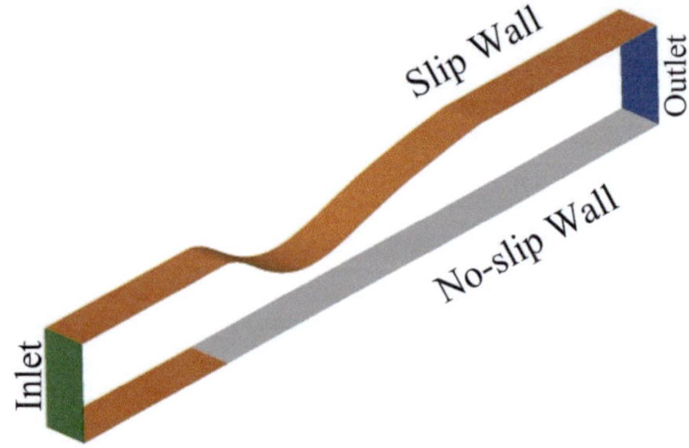

Fig. 5.8 L&H plate computational domain and local grid distribution

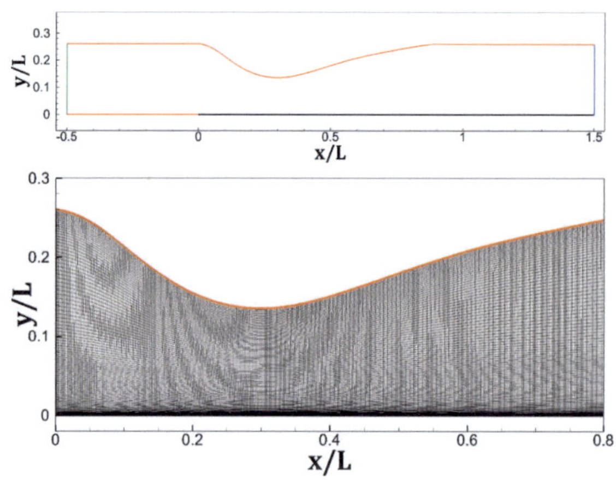

The specific computational parameters utilized in this case are outlined in Table 5.2. The Reynolds number, based on the plate length L and the inlet velocity U_{in}, is 60,000. The inlet turbulence intensity is 2.4%, while the turbulence intensity upstream of the separation bubble is 1.0%. In the case of unsteady simulations, a time step of 0.0005 L/U_{in} was employed. The simulation commenced with unsteady URANS results as the initial condition, which were then subjected to the unsteady $\gamma - A_r-$ VLES method. Following a simulation period of 15 L/U_{in}, data collection and processing for the mean flow field results was initiated, with a total simulation time of 25 L/U_{in}.

Table 5.2 Inlet boundary conditions for pressure gradient separation flat plate

Case	U_{in} (m/s)	Tu_{in} (%)	R_T	$\nu(\text{m}^2/\text{s}^{-1})$	Re
L&H tablet	3.4	2.4	3.0	1.5×10^{-5}	60,000

Figure 5.9 presents the distribution of surface pressure and friction coefficients on the plate, with a comparison of results from steady RANS, time-averaged $\gamma - A_r-$ VLES, and LES data from Lardeau et al. [46]. The $\gamma - A_r-$ VLES method predicts a smaller region with negative friction coefficients, indicating a smaller laminar separation bubble. This leads to differences in the surface pressure distribution when compared to LES data. In contrast, the $\gamma - A_r-$ VLES method aligns well with LES data for both friction and pressure coefficients. In addition, the LES results show a positive friction coefficient in the vicinity of $x/L = 0.5$, a detail that the $\gamma - A_r-$ VLES method accurately captures but that the RANS method fails to identify. The discrepancy in the predicted size of the separation bubble between the RANS and VLES methods results in variations in the pressure distribution across the plate, which in turn gives rise to discrepancies in the distribution of the surface friction coefficient prior to the onset of flow transition. For transitional flows that are characterized by the presence of significant separation bubbles and substantial unsteady effects, the capabilities of the RANS method to accurately simulate these flows are limited. Nevertheless, the $\gamma - A_r-$ VLES method, which has been developed based on a transition model, is able to effectively predict the location of the transition and to accurately describe the unsteady development of the separation bubble.

Figure 5.10 shows the distribution of turbulence intensity along the streamwise direction. The turbulence intensity predicted by the $\gamma - A_r-$ VLES method is in good agreement with the LES data from the inlet to the laminar separation bubble. In contrast to Lardeau et al.'s LES method, which utilizes synthetic turbulence at the inlet to introduce disturbances, the VLES method employs the RANS model prior to flow separation. The excellent alignment between these turbulence intensity profiles highlights the effectiveness of the boundary layer protection function F_{free} in the $\gamma - A_r-$ VLES method.

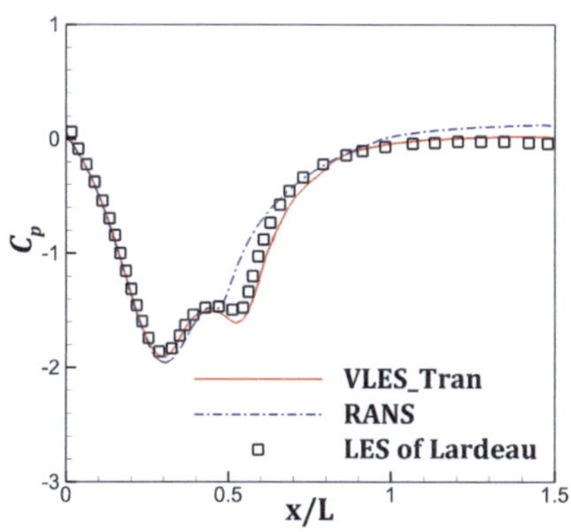

(a) Surface Pressure Coefficient Distribution

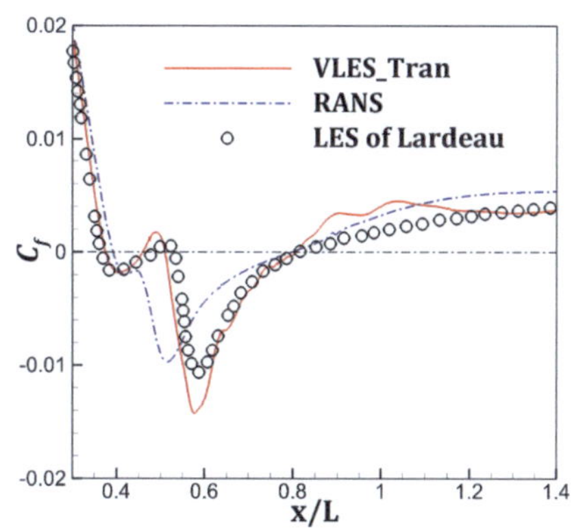

(b) Surface Friction Coefficient Distribution

Fig. 5.9 Distribution of surface pressure coefficient and friction coefficient for the L&H plate

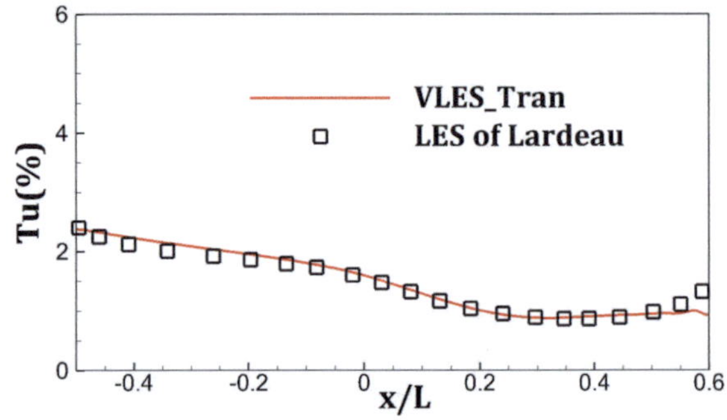

Fig. 5.10 Turbulence intensity distribution for the L&H plate

The transient evolution of the flow field computed using the $\gamma - A_r-$ VLES method for the flat plate case is illustrated over time. Figure 5.11 shows the distribution of the $Q(L/a_0)^2 = 0.005$ iso-surface, while Fig. 5.12 presents the spanwise vorticity distribution at the $z/L = 0.06$ plane. In the spanwise vorticity plot, at $x/L = 0.37$, the strong adverse pressure gradient causes the laminar boundary layer to separate, leading to the detachment and gradual lifting of the shear layer from the surface. As the flow progresses, around $x/L = 0.50$, periodic vortex shedding begins to develop, with the separated shear layer undergoing pronounced vertical oscillations, marking the onset of shear layer instability.

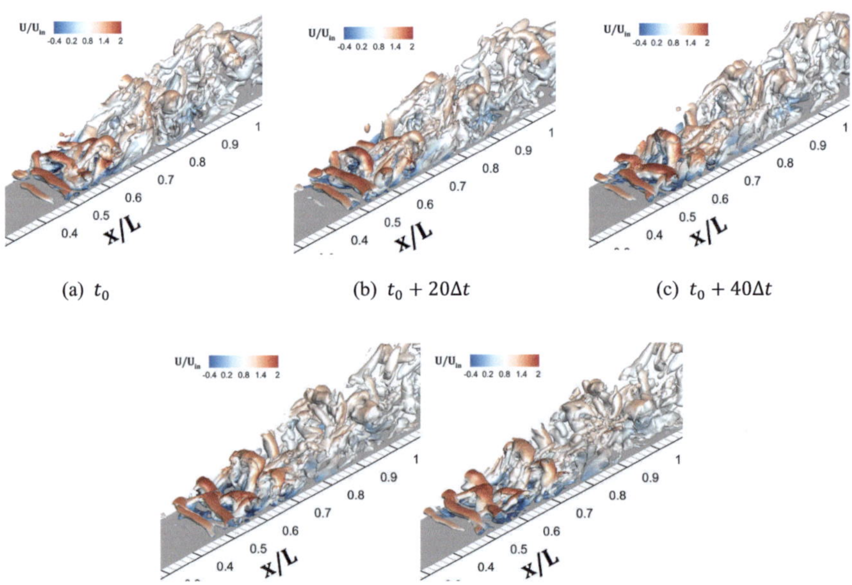

(a) t_0 (b) $t_0 + 20\Delta t$ (c) $t_0 + 40\Delta t$

(d) $t_0 + 60\Delta t$ (e) $t_0 + 80\Delta t$

Fig. 5.11 Transient flow field iso-surface distribution at $Q(L/a_0)^2 = 0.005$ simulated using the $\gamma - A_r-$ VLES method for the L&H flat plate

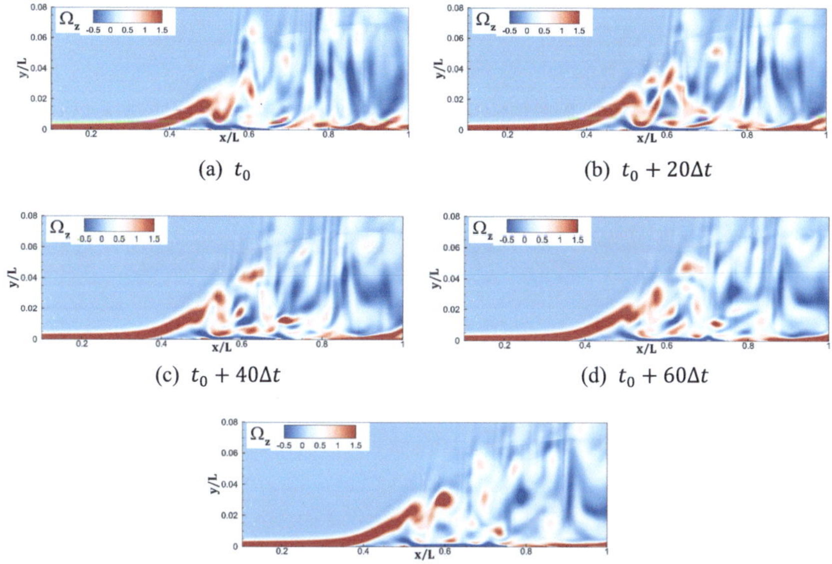

(a) t_0 (b) $t_0 + 20\Delta t$

(c) $t_0 + 40\Delta t$ (d) $t_0 + 60\Delta t$

(e) $t_0 + 80\Delta t$

Fig. 5.12 Spanwise vorticity distribution on the $z/L = 0.06$ plane of the transient flow field simulated using the $\gamma - A_r-$ VLES for the L&H flat plate

It is evident that the point at which the separated shear layer commences oscillatory behavior aligns with the onset of transition. This suggests that the instability of the separated shear layer (where transition begins) is closely linked to the oscillation of the separation bubble. The Q-criterion iso-surface plot also reveals that vortex shedding initiates around $x/L = 0.46$, gradually lifts off, and then breaks down, thereby initiating the transition. This leads to the formation of additional turbulent vortex structures, which in turn give rise to unsteady turbulent separation in the downstream region.

Figure 5.13 shows the mean streamwise velocity distribution in the vicinity of the laminar separation bubble, as determined by the $\gamma - A_r-$ VLES and RANS methods. Regions where the velocity is less than a certain threshold have been omitted from the presentation. The results are also compared with LES data from Lardeau et al. [46] (represented by black squares). While the RANS method yields a laminar separation bubble length that is in close agreement with the LES results, the bubble height is found to be smaller, and no secondary vortex structures are observed in the vicinity of the onset of transition.

In contrast, the $\gamma - A_r-$ VLES method predicts a larger separation bubble, with its height and size distribution exhibiting a high degree of correlation with the LES results. Additionally, the $\gamma - A_r-$ VLES method captures secondary vortices in the vicinity of the transition onset, which correlates with the observed increase in surface friction coefficient around $x/L = 0.5$ (see Fig. 5.9).

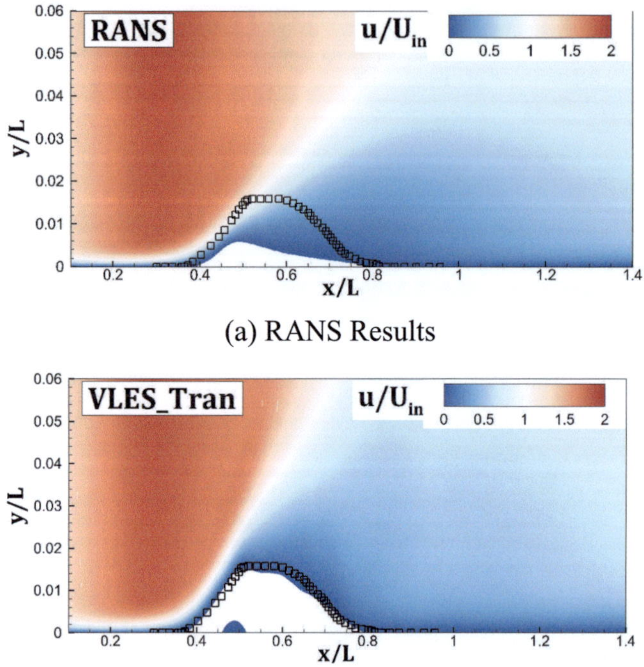

(b) VLES Mean Results

Fig. 5.13 Average velocity distribution near the laminar separation bubble on the L&H plate (Regions Where the Streamwise Velocity Is Less Than 0 Are Hidden; the Black Rectangle Indicates LES Data)

Figure 5.14 depicts the boundary layer velocity profiles at disparate streamwise locations on the flat plate. The $\gamma - A_r-$ VLES method's predictions for streamwise velocity exhibit a high degree of alignment with the LES data. However, there is a discernible tendency for the streamwise velocity to be slightly overestimated in the central region of the separation bubble. Notably, at $x/L = 0.5$, the $\gamma - A_r-$ VLES method predicts that the maximum reverse flow velocity occurs nearer to the wall compared to the LES results.

Figure 5.15 illustrates the distributions of Reynolds stresses at $x/L = 0.6$ and $x/L = 0.7$ within the boundary layer. The overall patterns of the Reynolds stresses predicted by the $\gamma - A_r-$ VLES method align well with the LES data, with both methods accurately capturing the locations of the stress extrema. Nonetheless, the $\gamma - A_r-$ VLES method exhibits a slight tendency to underestimate the peak values of $\overline{u'v'}$ in comparison to the LES results. In addition, at $x/L = 0.7$, the higher predicted separation bubble height and streamwise velocity in the $\gamma - A_r-$ VLES simulation leads to higher $\overline{u'u'}$ values than those observed in the LES data.

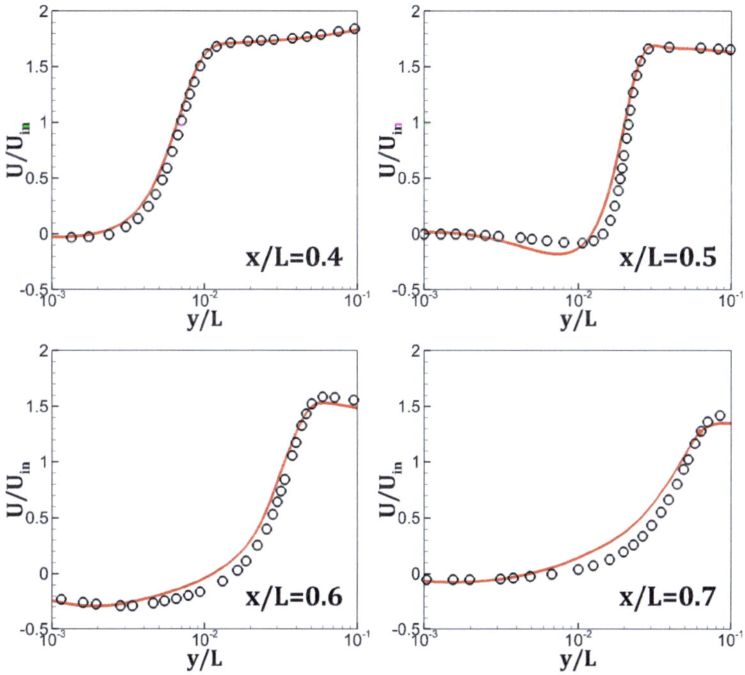

Fig. 5.14 Average velocity profiles at different streamwise positions on the L&H plate

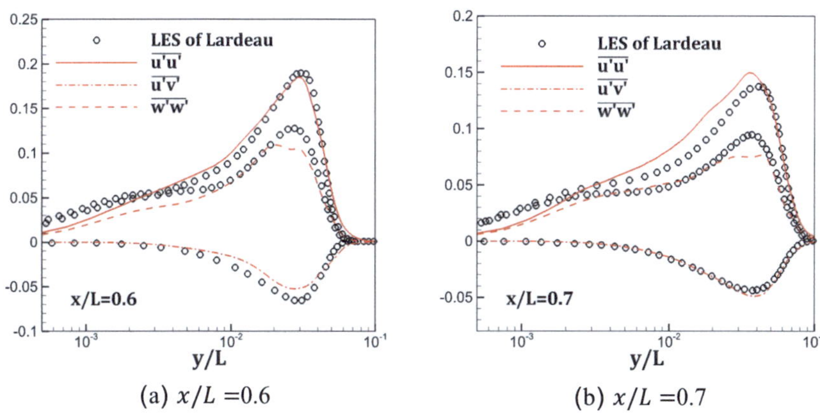

Fig. 5.15 Average reynolds stress distributions at different streamwise positions on the L&H plate

5.3.3 Flow Around Rough Cylinder

Achenbach [241] investigated the effect of surface roughness on flow around a cylinder through wind tunnel experiments. Utilizing a high-pressure wind tunnel and surface friction probes, he gathered data on the drag coefficient, surface friction coefficient distribution, pressure coefficient distribution, transition onset location, and boundary layer separation for different Reynolds numbers Re_D and roughness heights k_s. The cylinder model used in these experiments had a diameter of $D = 0.15$ m and a span of 0.5 m. Two different roughness levels were tested, with dimensionless equivalent sand-grain roughness heights k_s/D of 110×10^{-5} and 450×10^{-5}, respectively. The experiments were conducted in a low-speed pressurized wind tunnel, with the environment temperature held at 60 °C.

The experiments were conducted in a low-speed pressurized wind tunnel, with the ambient temperature maintained at 60 °C. Achenbach [241] observed that the dimensions of the test section (0.5 m × 0.9 m) were similar to the diameter of the cylinder, which introduced wall effects into the experimental data. Therefore, the computational domain dimensions were established in accordance with the positions of the experimental sections above and below the cylinder. The configuration of the computational domain is illustrated in Fig. 5.16a, with the grid distribution around the cylinder depicted in Fig. 5.16b, and the coordinate system on the cylinder surface illustrated in Fig. 5.16c.

Table 5.3 outlines the simulation conditions for a range of rough surfaces and Reynolds numbers. In this section, simulations are conducted to investigate the flow around cylinders with two distinct roughness levels, with each cylinder subjected to three different Reynolds numbers Re_D. The inlet turbulence intensity is matched to the experimental value, set at 0.7%. During the unsteady simulations, a time step of 1.37×10^{-5} s is applied. The simulations begin with unsteady URANS results as the initial field, followed by the addition of the unsteady VLES method. After an initial transient period of $15D/U_{\text{in}}$, the averaging of flow field results commences, with the total simulation time extending to $25D/U_{\text{in}}$.

Initially, the $\gamma - A_r$ transition model was used to conduct numerical simulations of the flow around rough cylinders under the RCY3 and RCY5 conditions (RANS TranRough). Simultaneously, the γ transition model was employed to simulate the flow around a smooth cylinder under identical inflow conditions (RANS TranSmooth). The resulting distributions of surface friction coefficients for the cylinders are presented in Fig. 5.17.

In the case of the smooth cylinder, the boundary layer remains laminar up to the point of separation at both Reynolds numbers. As a result, the γ model forecasts a gradual change in the surface friction coefficient preceding the point of separation. In contrast, for the rough cylinder, the boundary layer undergoes transition before separation. The $\gamma - A_r$ model predicts a transition point that closely aligns with the experimental data reported by Achenbach [241]. These findings underscore the substantial impact of surface roughness on the evolution of the boundary layer and the transition process in the vicinity of the cylinder. It is therefore essential that the

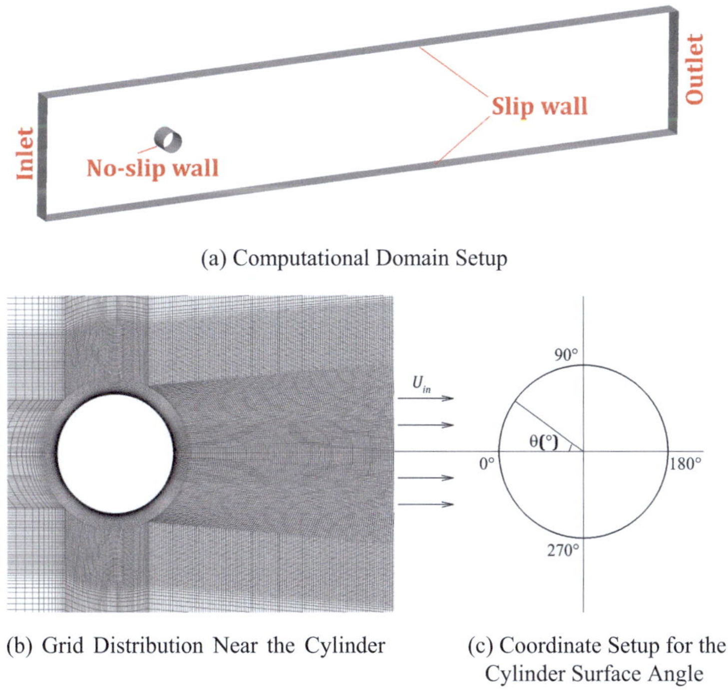

Fig. 5.16 Computational domain and grid distribution near the cylinder, and coordinate setup for the cylinder surface

Table 5.3 Surface settings and simulation conditions for rough cylinders

Case	k_s/D	Re_D	Tu (%)
RCY1	110×10^{-5}	13×10^4	0.7
RCY2	110×10^{-5}	43×10^4	0.7
RCY3	110×10^{-5}	65×10^4	0.7
RCY4	110×10^{-5}	10×10^4	0.7
RCY5	110×10^{-5}	17×10^4	0.7
RCY6	110×10^{-5}	300×10^4	0.7

effects of surface roughness are incorporated when simulating the flow around a rough cylinder. The subsequent analysis will not include the simulation results for smooth cylinders, as the focus is on rough cylinders.

In order to provide a comprehensive evaluation of the simulation precision of the novel hybrid approach in comparison to the RANS method, numerical simulations of flow around rough cylinders were conducted utilizing four distinct techniques: the $k - \omega$ SST turbulence model (RANS TurbRough), the $\gamma - A_r$ transition model (RANS TranRough), the fully turbulent VLES hybrid method (VLES TurbRough), and the $\gamma - A_r-$ VLES hybrid method (VLES TranRough). The objective of these

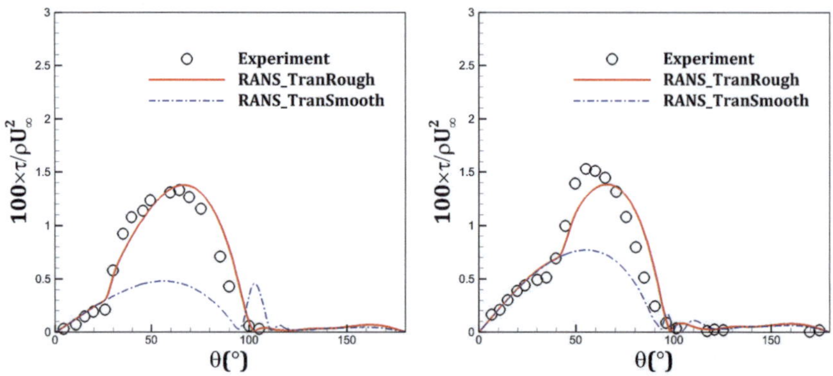

(a) RCY3 Status (b) RCY5 Status

Fig. 5.17 Distribution of cylinder surface friction coefficient predicted by the RANS transition model

simulations was to investigate the transition process in the boundary layer and the development of wake vortices under a range of Reynolds numbers and surface roughness conditions. In addition, the results were also compared with simulations based on the RANS roughness transition model proposed by Son et al. [214, 242] and experimental data from Achenbach [241]. The distributions of the time-averaged and spanwise-averaged surface friction coefficient (C_f) for each computational state are illustrated in Fig. 5.18. The $\gamma - A_r$ TranRough and VLES TranRough methods yielded very similar C_f distributions, and both showed better agreement with experimental data compared to the results from Son et al. This was evident in their ability to predict the location of the transition, defined as the point at which C_f exhibits a pronounced increase.

At the cylinder's separation point (where C_f is negative), the predictions provided by VLES TurbRough model are more consistent with experimental results compared to other methods. In the RCY1 case, the C_f distribution reveals that all numerical methods predict that the boundary layer on the cylinder will remain laminar up to the point of separation. This results in the laminar separation of the boundary layer, which is in accordance with the findings of the experimental investigation. Nevertheless, all methods indicate that the separation point will occur at a later time than observed in the experiments.

The distribution of the time-averaged and spanwise-averaged surface pressure coefficient for the cylinder is depicted in Fig. 5.19. In the RCY1 case, experimental measurements show that the laminar separation of the cylinder's boundary layer occurs around $\theta = 70°$, whereas all the models predict the separation point near $\theta = 90°$. It indicates a clear discrepancy between the numerical simulations and the experimental results. In other cases, the VLES TranRough model's predictions for both the peak and plateau pressure coefficient values are more closely aligned with the experimental data compared to the other models.

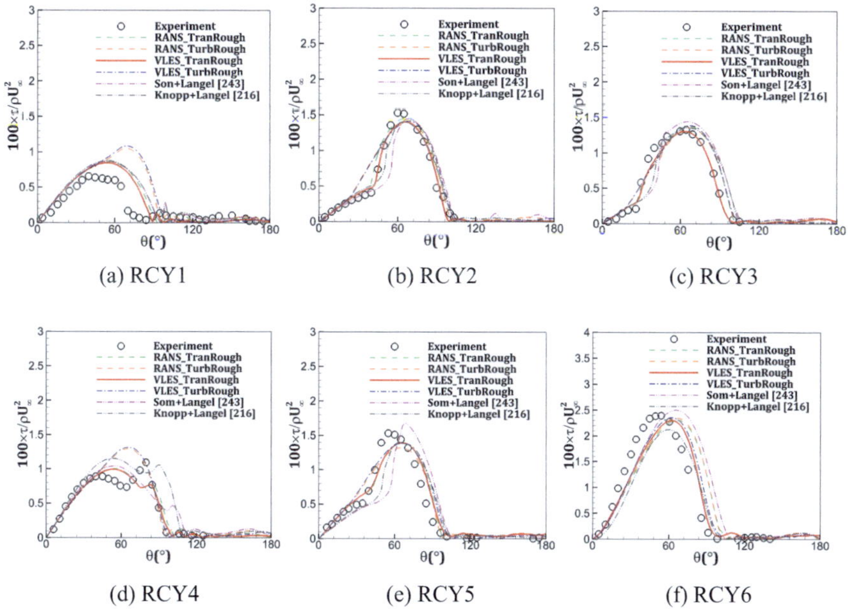

(a) RCY1 (b) RCY2 (c) RCY3

(d) RCY4 (e) RCY5 (f) RCY6

Fig. 5.18 Distribution of mean surface friction coefficient on a rough cylinder

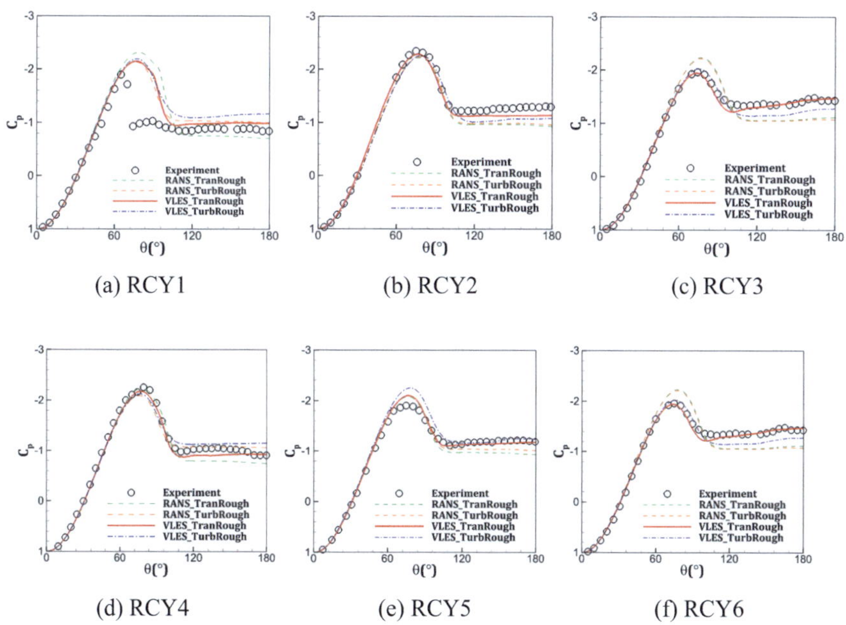

(a) RCY1 (b) RCY2 (c) RCY3

(d) RCY4 (e) RCY5 (f) RCY6

Fig. 5.19 Distribution of mean pressure coefficient on a rough cylinder

The boundary layer transition onset and wake separation positions for the rough cylinder under various computational conditions are illustrated based on numerical simulation results and compared with findings from Achenbach [241] and Son et al. [214, 242], as shown in Figs. 5.20 and 5.21. As full turbulence models are unable to predict the transition process, their results are only included in the comparison of separation positions. The transition position analysis indicates that the RANS TranRough and VLES TranRough models predict transition locations that are closely aligned with each other and are in good agreement with experimental results, whereas Son et al.'s predictions place the transition point further downstream. In the separation position comparison, the VLES TranRough model's predicted separation onset is consistent with experimental data, while the predictions from other methods show considerable deviations. These findings demonstrate that the $\gamma - A_r -$ VLES method offers improved accuracy in simulating rough cylinder flow compared to other methods.

Figure 5.22 presents a comparison between the mean drag coefficients predicted by various methods and the corresponding experimental data [241]. As the Reynolds number increases, the boundary layer on the cylinder transitions from laminar to

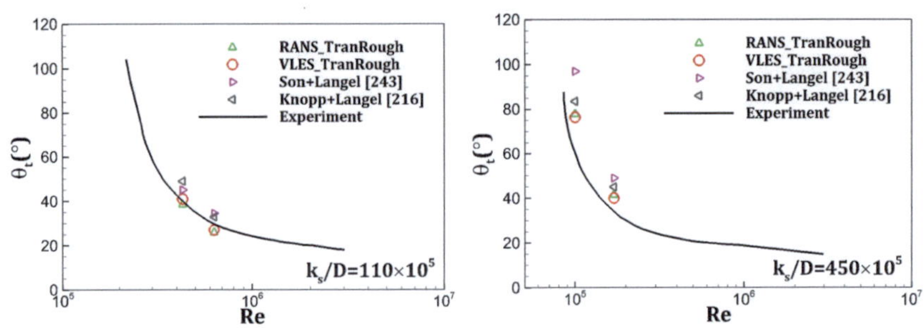

(a) Surface Roughness k_s/D =110×10^{-5} (b) Surface Roughness k_s/D=450×10^{-5}

Fig. 5.20 Comparison of boundary layer transition onset positions for the rough cylinder

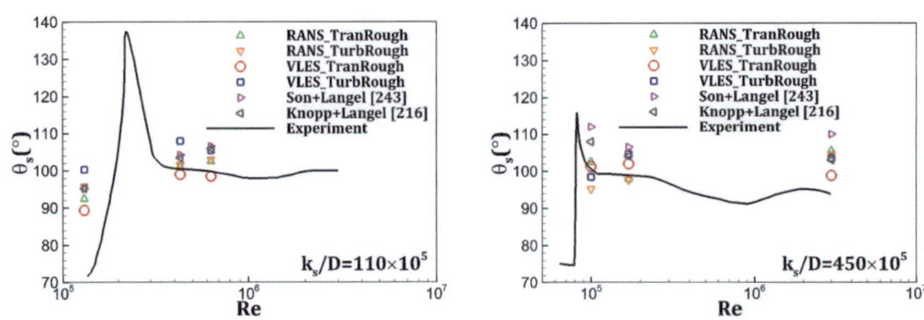

(a) Surface Roughness k_s/D=110×10^{-5} (b) Surface Roughness k_s/D=450×10^{-5}

Fig. 5.21 Comparison of wake separation positions for the rough cylinder turbulent, resulting in the drag coefficient to first decrease and then increase-a phenomenon known as the "drag crisis". The experimental data demonstrates this behavior in the rough cylinder, where an increase in surface roughness results in a reduction of the critical Reynolds number and a subsequent elevation of the drag coefficient beyond the threshold. Both the RANS TranRough and VLES TranRough models are able to successfully capture this variation in drag. In contrast, full turbulence models, which are unable to account for boundary layer transition, are unable to reflect this trend. In comparison to RANS TranRough, VLES TranRough not only predicts a higher drag coefficient but also demonstrates superior agreement with the experimental data.

Figure 5.23 presents the distribution of the transient intermittency factor within the boundary layer in the vicinity of the cylinder wall, as predicted by the $\gamma - A_r-$ VLES method. As the Reynolds number increases for a cylinder with a uniform rough surface, the location at which the intermittency factor reaches 1.0 shifts upstream, indicating an earlier transition. In the RCY6 case, the intermittency factor in the vicinity of the cylinder surface is observed to be consistently equal to one, which indicates that the boundary layer in this region is predominantly turbulent.

Fig. 5.22 Comparison of mean drag coefficient for the rough cylinder

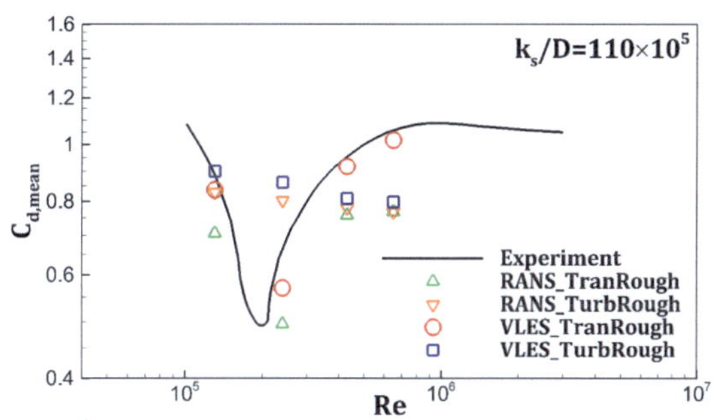

(a) Surface Roughness k_s/D =110×10^{-5}

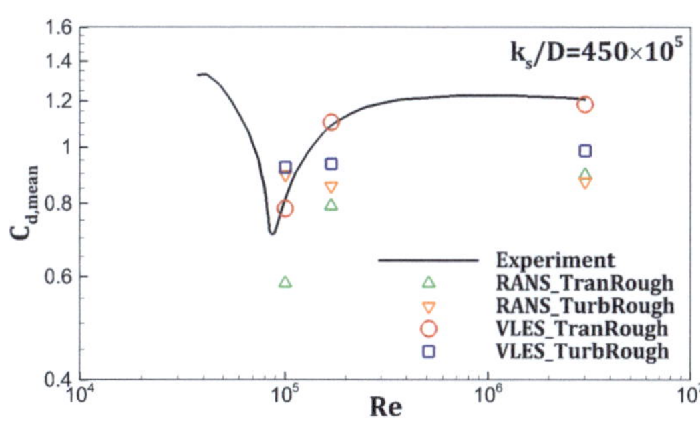

(b) Surface Roughness k_s/D=450×10^{-5}

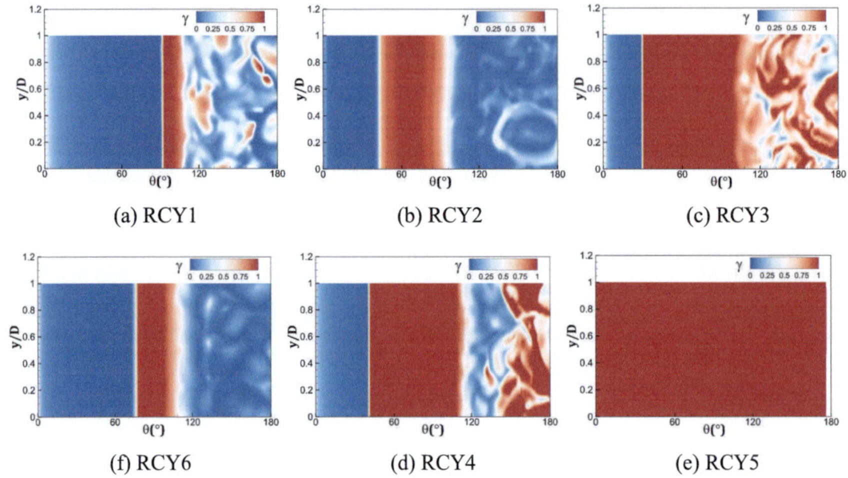

Fig. 5.23 Distribution of transient intermittency factor γ near the surface of the rough cylinder predicted by the $\gamma - A_r -$ VLES method

Figure 5.24 shows the distribution of the Q criterion iso-surfaces for the transient flow field of the rough cylinder, computed using the $\gamma - A_r -$ VLES method. The Q criterion plots are colored according to the stream wise velocity. The figure reveals that, in all states, the cylinder's wake is turbulent and features alternating vortex shedding with significant three-dimensional effects. For cylinders with the same rough surface, as the Reynolds number increases, the location of the shear layer instability in the wake progressively moves upstream.

Simulations were performed on a high-performance computing cluster. When the same conditions were applied, including the same computer, number of CPUs, and simulation time steps, the $\gamma - A_r -$ VLES method required 30% more computational time than the SST-VLES method. The additional computational time can be attributed to the two additional transport equations in the $\gamma - A_r -$ VLES method, which significantly increase the computational load. The observed increase in computation time aligns closely with the ratio reported by Hodara et al. [230].

5.3.4 Flow Around the Rough V103 Compressor Cascade

The internal flow environment within a compressor is characterized by a high degree of intricacy, which is influenced by a number of factors, including the leading-edge shape of the blade, the curvature of the suction surface, the presence of strong pressure gradients, and the properties of the free-stream flow that exists beyond the boundary layer [243]. At lower Reynolds numbers, high-load blades are susceptible to laminar separation on the suction side. Subsequently, the separated shear layer typically undergoes a transition to turbulence, which results in rapid reattachment upstream of the trailing edge. This results in the formation of a separation bubble,

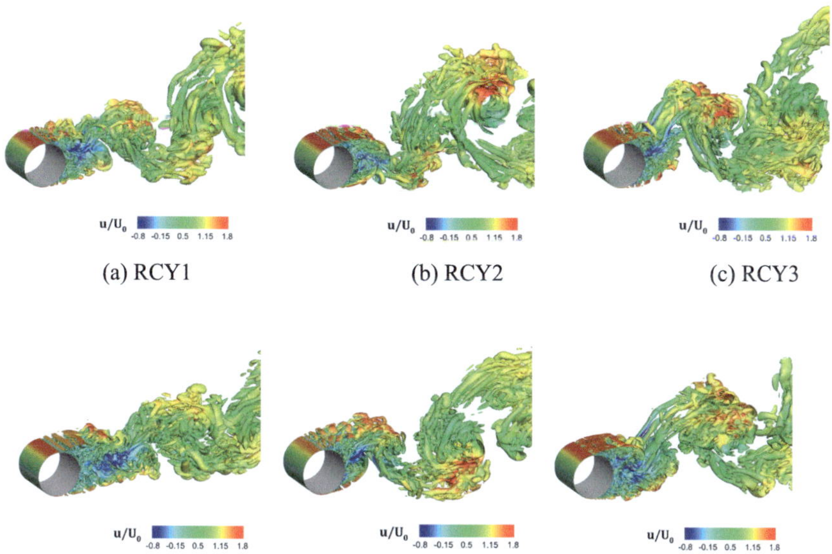

(a) RCY1 (b) RCY2 (c) RCY3

(d) RCY4 (e) RCY5 (f) RCY6

Fig. 5.24 Iso-surface distribution of $Q(L/a_0)^2 = 0.1$ in the transient flow field of the rough cylinder predicted by the $\gamma - A_r-$ VLES method

which alters the effective blade profile, thereby reducing compressor efficiency and increasing both profile and wake losses [244]. What's more, issues such as combustion deposits, corrosion and material erosion can alter the blade's surface geometry, increasing roughness and influencing the behavior of boundary layer separation and transition. It is therefore imperative to accurately model the impact of surface roughness on separation bubble-induced transitional flow in order to gain a deeper understanding of turbomachinery performance. This section presents a numerical simulation of the flow around the rough V103 compressor blade, evaluating the $\gamma - A_r-$ VLES method's effectiveness in capturing the complex flow dynamics within turbomachinery.

In their study, Wang et al. [245] employed large eddy simulation (LES) to examine the impact of distributed surface roughness on the separated flow field of the high-load V103 compressor blade. The roughness elements were positioned downstream of the suction surface's velocity peak, with four different roughness sizes considered, corresponding to roughness heights k_s of 0, 72, 144, and 288 μm. In their LES-based flow field simulations, Wang et al. incorporated a geometric modeling approach for the roughness elements, defining the surface roughness height y_{wall} as follows [245]:

$$y_{\text{wall}} = \max(0, k\sin(k_x x) \cdot \sin(k_z z)), \text{ if } x > 0.22C_{ax} \tag{5.27}$$

$$y_{\text{wall}} = 0, \text{ elsewhere.} \tag{5.28}$$

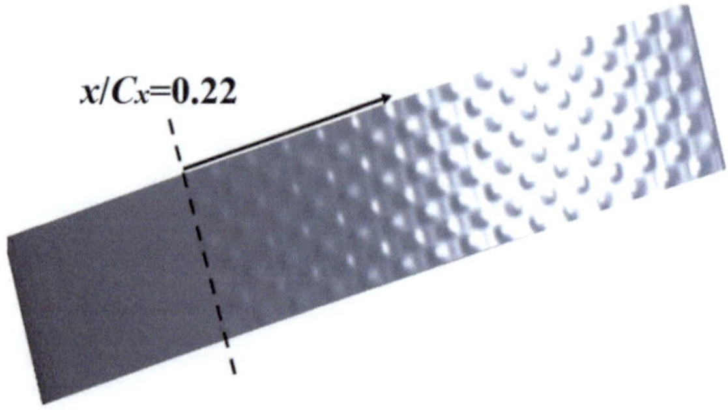

Fig. 5.25 Schematic of the rough blade surface of the V103 compressor simulated by Wang et al. using LES [245]

In the numerical simulations conducted by Wang et al., when $k_s = 288\ \mu m$, the corresponding geometric shape of the rough blade surface is shown in Fig. 5.25 [245].

In this study, the $\gamma - A_r -$ VLES method is employed to simulate the flow around the V103 compressor blade with four distinct rough surface configurations. The V103 compressor model is depicted in Fig. 5.26, and the relevant geometric parameters and simulation conditions are outlined in Table 5.4. In lieu of a geometrical modeling of the roughness elements on the blade surface, the simulation employs an equivalent roughness height to define the wall roughness growth factor, thereby capturing the effects of surface roughness on flow separation and transition. For $x > 0.22C_x$, the suction side of the blade is treated as a rough wall, with the models designated as "Smooth," "Rough1," "Rough2," and "Rough3," corresponding to roughness heights of 0, 72, 144, and 288 μm, respectively.

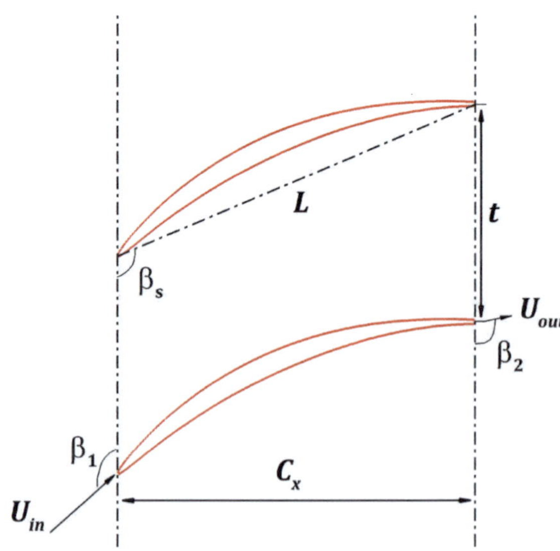

Fig. 5.26 Schematic diagram of the V103 compressor model

Table 5.4 Geometric parameters and simulation conditions for the V103 compressor

Chord length L(mm)	180
Axial chord length C_x(mm)	166.3
Pitch,t(mm)	99
Inflow angle,β_1(°)	132
Outflow angle,β_2(°)	96
Installation angle,β_s(°)	112.5
Inlet match number, Ma	0.67
Inlet Reynolds number,Re_L	1.5×10^5
Inlet turbulence intensity,Tu_{in}	3.5%

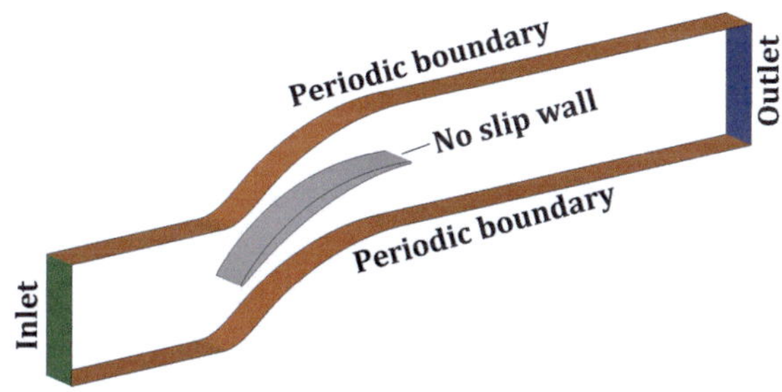

Fig. 5.27 The boundary settings for the V103 compressor computational domain

In this instance, the computational domain boundaries are defined as illustrated in Fig. 5.27. The spanwise boundary is specified as a periodic boundary condition with a spanwise length of 0.2 C_x. Figure 5.28 illustrates the dimensions of the computational domain and the distribution of the mesh in the vicinity of the separation bubble on the blade suction surface. The inlet boundary is set at $x/C_x = -1$, and to accurately simulate the development of the compressor wake, the outlet boundary is set at $x/C_x = -3$. The total number of mesh cells for the entire flow field is approximately 7.2 million. In the region around the separation bubble, the maximum mesh resolutions in the streamwise, normal, and spanwise directions are designed to meet $\Delta x^+ < 15$, $\Delta y^+ < 1$ and $\Delta z^+ < 15$, respectively.

In the unsteady mixed-method simulation, the time step is 1.0×10^{-6} S. The VLES method is initially applied for a duration of $15L/U_{in}$ to achieve a fully developed unsteady flow field within the compressor. Subsequently, the flow field results are averaged over a period of $25L/U_{\mathrm{in}}$ through both time and spanwise averaging to produce the mean flow field. This averaged result is then compared and analyzed against the LES data [245].

The simulation of flow around the smooth surface of the compressor blade is conducted initially. The isentropic Mach number distribution along the mean surface of the blade, as depicted in Fig. 5.29, reveals that the $\gamma - A_r -$ VLES method (VLES Tran) simulation outcomes exhibit a high degree of correlation with both the experimental data [246] and the LES results. A significant plateau region is observed on the suction side of the blade, suggesting that the laminar boundary layer separation and

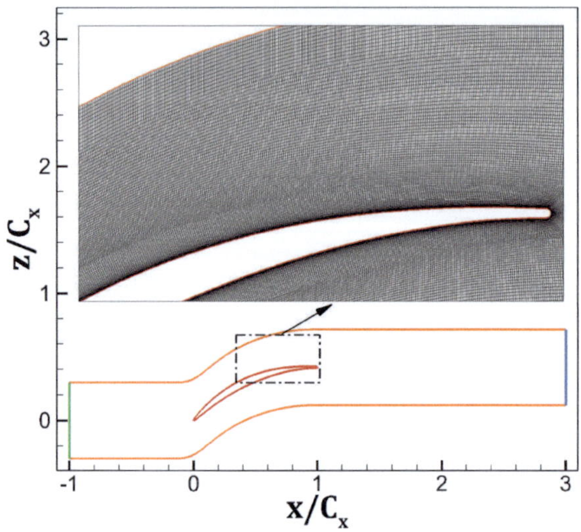

Fig. 5.28 Computational domain dimensions and local mesh distribution for the V103 compressor

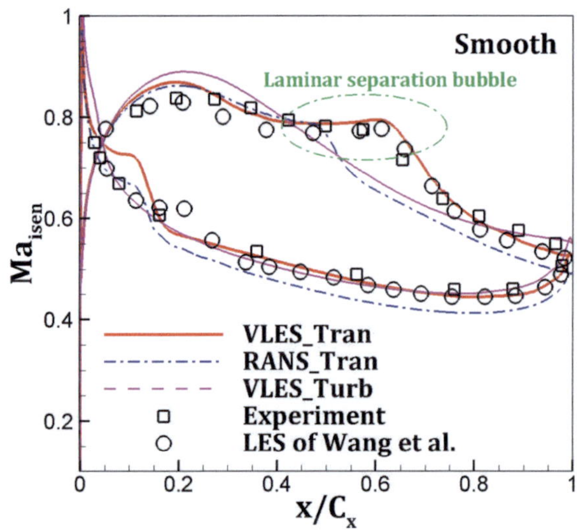

Fig. 5.29 Comparison of isentropic mach number on the average surface of the smooth V103 compressor

the subsequent transition process have resulted in the formation of a closed laminar separation bubble on the blade surface.

In contrast, the isentropic Mach number distribution predicted by the $\gamma - A_r$ transition model (RANS Tran) shows significant deviations from the experimental results and LES data. The RANS method forecasts a markedly shorter plateau region and a smaller laminar separation bubble.

Figure 5.30 presents the distribution of the average surface friction coefficient on the suction side of the smooth V103 compressor blade. The figure includes the results obtained from the $\gamma - A_r$ transition model (RANS Tran), the $\gamma - \overline{Re_{\theta t}}$ transition model

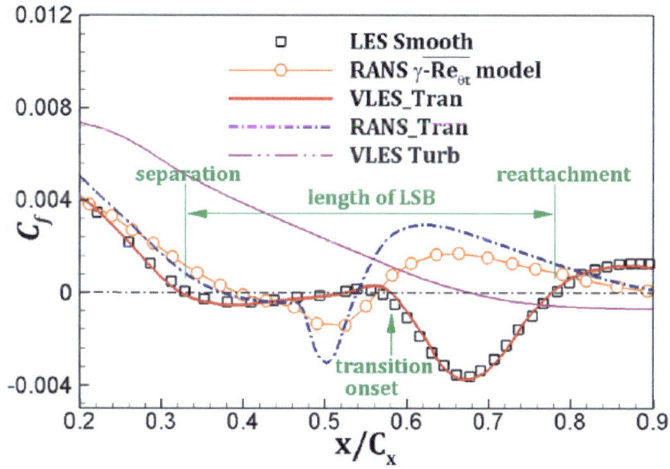

Fig. 5.30 Comparison of average surface friction coefficient on the smooth V103 compressor

(RANS $\gamma - \overline{Re_{\theta t}}$ model), the $\gamma - A_r-$ VLES method (VLES Tran), the $\gamma - A_r-$ VLES method (VLES Turb), and the LES method [245] (LES Smooth). Furthermore, it indicates the separation points, transition points, and reattachment points predicted by the LES method. The VLES method predicts the location of separation in close proximity to the blade's trailing edge and fails to capture the central separation bubble. In contrast, the RANS method is unable to accurately capture the unsteady characteristics of the laminar separation bubble. In comparison to the LES data, the RANS method results in a delay in the prediction of laminar separation and an advancement in the position of turbulent reattachment, which in turn leads to a significantly smaller laminar separation bubble. Similarly, Wang et al. [247] observed comparable discrepancies in their comparison of RANS and LES simulations of blade flow. The $\gamma - A_r-$ VLES method exhibits a high degree of concordance with the LES data, with the positions of separation, transition, and reattachment aligning closely with those observed in the LES. Furthermore, the method effectively captures the secondary vortex structures that form near the transition location within the separation bubble.

Figure 5.31 presents the distribution of the average surface friction coefficient for the V103 compressor with varying roughness configurations. The comparison between the predictions from the $\gamma - A_r-$ VLES method and the LES method is presented. Prior to the transition, both methods exhibit a good agreement in the C_f distribution. Following the transition period, the C_f value declines rapidly in both methodologies, reflecting comparable trends, although there are slight discrepancies in the distribution profiles. The predicted separation points exhibit a high degree of alignment between the two methods, and the location of separation demonstrates a notable degree of stability as the roughness height increases. In comparison to the LES data, the $\gamma - A_r-$ VLES method predicts a marginally downstream position for

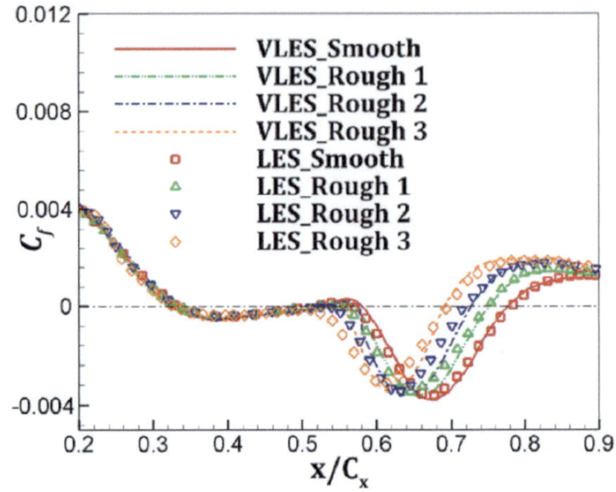

Fig. 5.31 Comparison of average surface friction coefficient for the V103 compressor with different rough surfaces

turbulent reattachment, yet both methods indicate an upstream shift in the reattachment location with increasing roughness height. At approximately $x/C_x = 0.55$, the $\gamma - A_r -$ VLES method predicts a positive surface friction coefficient, suggesting the presence of secondary vortices and denoting the transition location. Conversely, the LES method forecasts a reduction in the $C_f > 0$ region with increasing roughness height, with no such region remaining on the Rough3 roughened blade. The $\gamma - A_r -$ VLES method predicts a slightly larger secondary vortex size relative to the LES results, and the upstream shift of the transition location with increased roughness height is less pronounced. Even for the Rough3 roughened blade, the $\gamma - A_r -$ VLES method still predicts a $C_f > 0$ region around $x/C_x = 0.55$, which leads to a downstream delay in the rapid decrease of C_f compared to the LES results.

A further comparison of the mean velocity profiles at varying streamwise locations within the separation bubble is presented in Fig. 5.32. At $x/C_x = 0.45$, the boundary layer is already separated and lies within the laminar region of the separation bubble. As the roughness height increases, the $\gamma - A_r -$ VLES method predicts a reduction in the maximum reverse flow velocity and an increase in the boundary layer edge velocity. This trend is in accordance with the results obtained from the LES.

In the case of blades exhibiting four distinct surface roughness levels, the velocity profiles predicted by the VLES method demonstrate a high degree of correlation with the LES data. At $x/C_x = 0.55$, an increase in surface roughness is observed to result in a reduction in the height of the separation bubble. The $\gamma - A_r -$ VLES method predicts the formation of a larger secondary vortex than is observed in the LES results. This leads to a reduction in the maximum reverse flow velocity in the VLES predictions at this location. Despite this, the velocity profiles at other streamwise positions are relatively consistent between the two methods. At the latter two streamwise locations, the $\gamma - A_r -$ VLES results align well with the LES data. At $x/C_x = 0.70$, where the boundary layer on the suction side of the Rough3 roughened blade has reattached, both methods predict positive velocity distributions, which are indicative of the characteristics of a turbulent boundary layer.

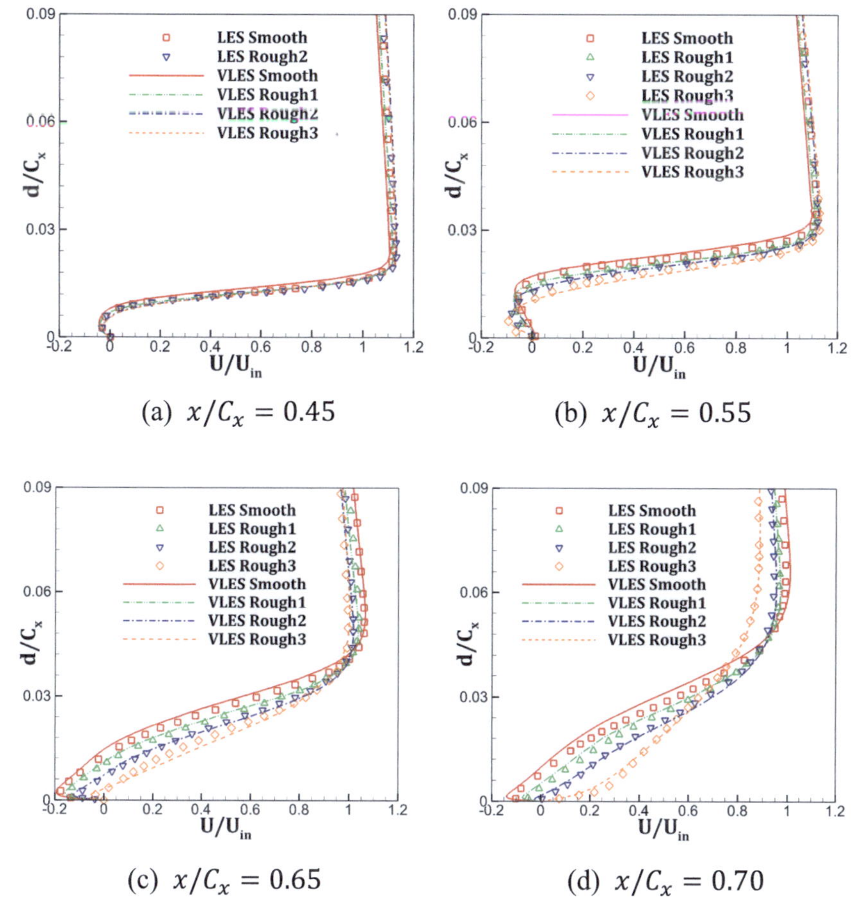

Fig. 5.32 Comparison of velocity profiles at various streamwise locations for the V103 compressor with different surface roughness levels

Figures 5.33 and 5.34 present a comparison of the mean Reynolds stresses in the boundary layer at $x/C_x = 0.75$ and $x/C_x = 0.85$. These positions are in close proximity to the reattachment point of the separation bubble. The overall distribution patterns predicted by the $\gamma - A_r -$ VLES method closely match the LES data, with the height at which the Reynolds stresses peak being nearly identical. However, because the $\gamma - A_r -$ VLES method predicts a slightly larger separation bubble than the LES results, the Reynolds stresses predicted at these positions are also slightly higher than those in the LES data.

Figure 5.35 shows the distribution of spanwise vorticity Ω_y at the $y/C_x = 0.75$ plane in the transient flow field of the V103 compressor with different surface roughness levels, as predicted by the VLES method. The figure reveals the presence of

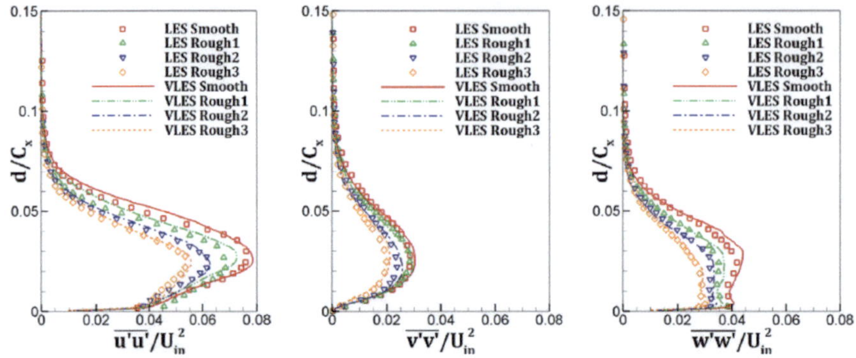

Fig. 5.33 Comparison of reynolds stresses at $x/C_x = 0.75$ for the V103 compressor with different surface roughness levels

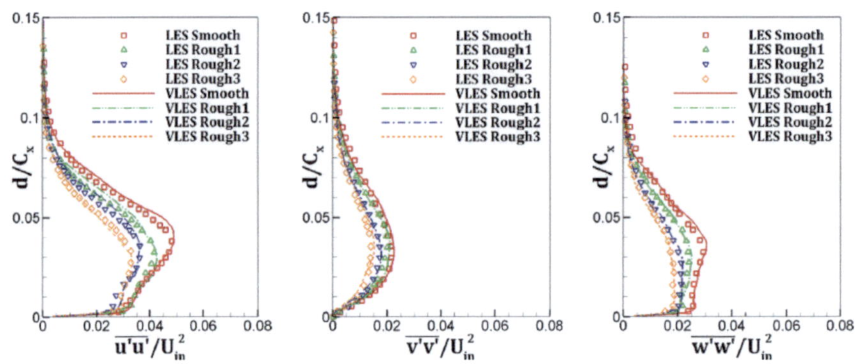

Fig. 5.34 Comparison of reynolds stresses at $x/C_x = 0.85$ for the V103 compressor with different surface roughness levels

unsteady separation bubbles on both sides of the blade. An increase in surface roughness on the suction side of the blade results in a discernible intensification of the disturbances caused by the roughness elements. This results in an upstream movement of the instability point in the separation bubble's shear layer, which in turn causes the transition point to move forward. Concurrently, the region affected by the unsteady separated flow after the transition undergoes a gradual reduction in size, accompanied by a decrease in the dimensions of the separation bubble. The $\gamma - A_r-$ VLES method accurately captures this trend and effectively reflects the impact of surface roughness on boundary layer transition and the development of separation vortices.

Figure 5.36 presents the Q-criterion iso-surface distribution in the transient flow field on the suction side of blades with varying surface roughness, as predicted by the $\gamma - A_r-$ VLES method. The subfigures, spaced 20-time steps apart and color-coded

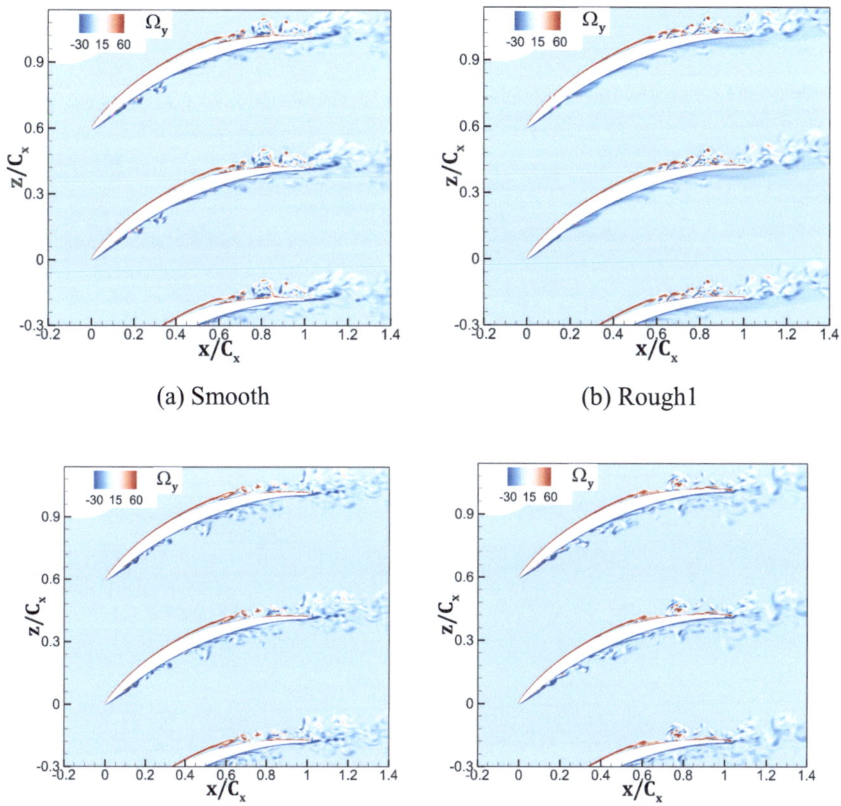

(a) Smooth (b) Rough1

(c) Rough2 (d) Rough3

Fig. 5.35 Distribution of spanwise vorticity Ω_y at $y/C_x = 0.1$ in the transient flow field of the V103 compressor with different surface roughness levels

based on the X-axis velocity, depict the periodic shedding and evolution of separation vortices from the shear layer on the suction sides of the "Smooth," "Rough1," "Rough2," and "Rough3" blades as the simulation progresses. In Fig. 5.36a, it is evident that, influenced by the adverse pressure gradient and blade curvature, laminar separation occurs near $x/C_x = 0.33$. At the outset, two-dimensional spanwise vortices are shed from the upstream separation shear layer. As the separation bubble expands, the vortices are lifted and deformed, gradually exhibiting three-dimensional motion. Over time, increasing disturbances in the separation shear layer stretch the spanwise vortices, which then evolve into streamwise vortices. The interactions between these vortices result in accelerated deformation, which ultimately leads to the breakdown of the vortices into smaller-scale turbulent vortices. An increase in surface roughness height results in a relatively constant laminar separation point within the boundary layer, but a significant influence on the transition

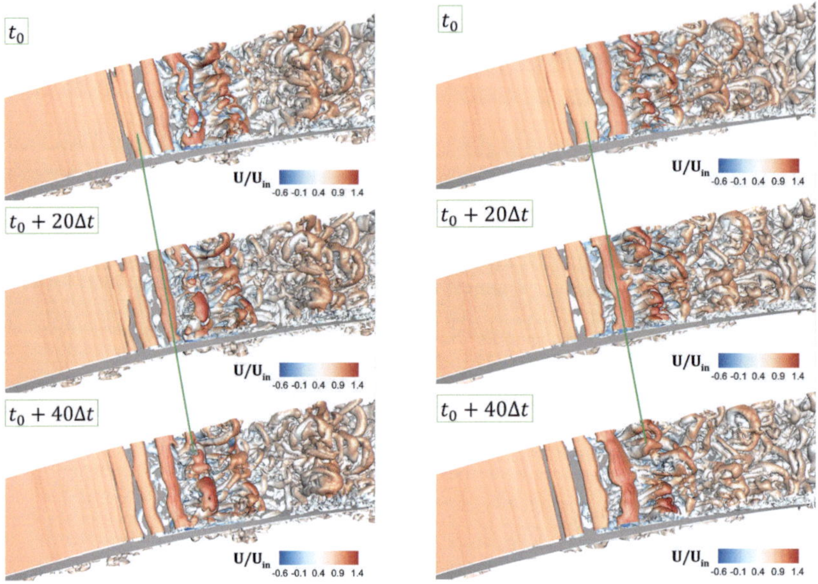

(a)Smooth (b) Rough1

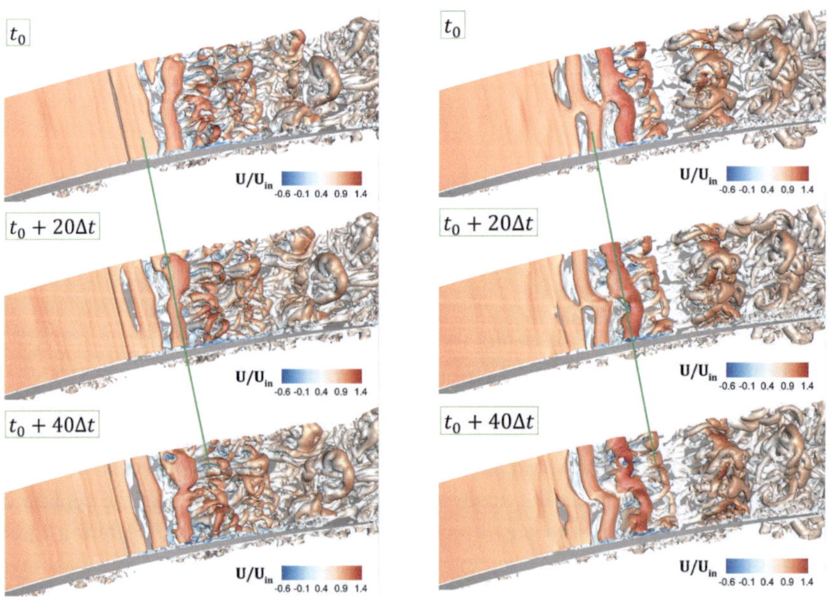

(c)Rough2 (d) Rough3

Fig. 5.36 Iso-surface distribution of $Q(C_x/a_0)^2 = 0.1$ on the suction side of the V103 compressor with different surface roughness levels process and vortex dynamics within the separation bubble. Although the shedding and evolution of vortices follow a pattern similar to that observed on a smooth blade, the enhanced surface roughness height intensifies disturbances within the separation shear layer at corresponding streamwise positions. This results in the onset of three-dimensional motion in the spanwise vortices occurring at an earlier point in time, with streamwise vortices forming at an earlier stage and breaking down into small-scale turbulent vortices at a more rapid rate. This evolution of vortices is in close alignment with the results observed in LES simulations [245].

5.3.5 *Flow Around the Smooth Pak-B Turbine Blade*

As airflow passes over the low-pressure turbine blades of an aircraft engine, the high loading on the blades gives rise to a tendency for laminar boundary layer separation bubble transition on the suction side. If further separation with highly unsteady characteristics occurs, it can alter the blade's effective shape and degrade performance. In order to assess the predictive accuracy of the previously introduced LES method, which accounts for transition, simulations and analyses of the flow around the turbine blades are carried out, with a particular focus on the prediction of separation bubble transition on the surface of low-pressure turbine blades.

The Pak-B low-pressure turbine blade was designed by Pratt and Whitney, features loading conditions similar to those of modern aircraft engine rotors [248]. These conditions represent typical low-pressure turbine operation at low Reynolds numbers during high-altitude cruising. The transition occurs within the free shear layer of the separation bubble on the suction side; it influences both the reattachment of the separation bubble and the stall characteristics of the blade. A schematic of the Pak-B turbine blade geometry and its specific parameters are presented in Table 5.5 and Fig. 5.37 for reference. Simulations of the Pak-B blade were conducted using the γ − VLES model introduced in this study, with results compared against the r-transition model and $\gamma - \overline{Re_{\theta t}}$ transition model, both of which are based on the RANS method. The experiment was performed at a Reynolds number of 50,000 with an inlet turbulence intensity (Tu_{∞}) of 0.08%. The boundary conditions for the computational domain are illustrated in Fig. 5.38, where C_x denotes the chord length of the low-pressure turbine blade. Periodic boundary conditions were applied in the spanwise direction, with the spanwise length set to $0.2C_x$. The inlet boundary was

Table 5.5 Geometry parameters of Pak-B

Axial chord (C_x)/mm	75
True chord (C)/mm	83.4
Blade spacing (S)/mm	66.8
Stagger angle $(\gamma)/(°)$	26
Inlet angle $(\beta_1)/(°)$	35
Outlet angle $(\beta_2)/(°)$	60

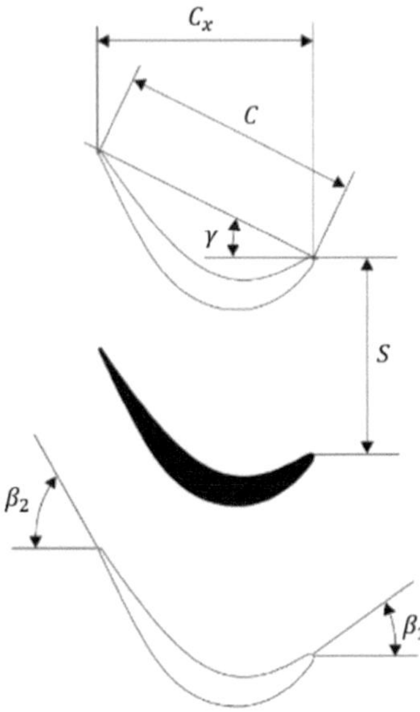

Fig. 5.37 Pak-B turbine blade geometry

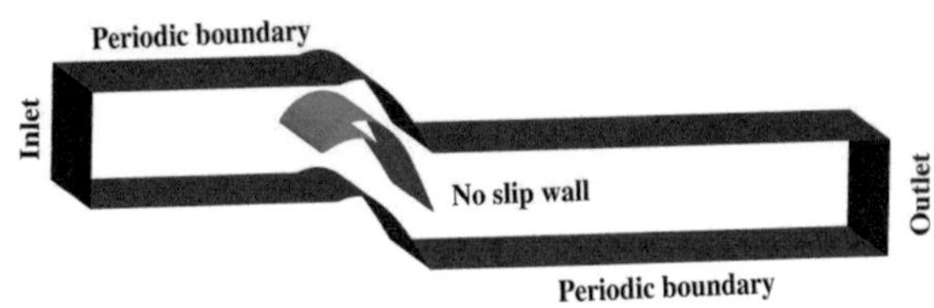

Fig. 5.38 Pak-B turbine blade computational domain boundary settings

set at $-2C_x$, with the outlet boundary located at $6C_x$, both with respect to the blade's leading edge.

As illustrated in Fig. 5.39, a grid sensitivity analysis was conducted for the Pak-B turbine blade utilizing three different mesh resolutions, with the details of each grid provided in Table 5.6. For all grids, the non-dimensional wall distance of the first layer of cells on the blade surface is maintained at $y^+ < 1$. The γ − VLES method was employed for the simulations, and the skin friction coefficient along the blade's upper surface for each grid is depicted in Fig. 5.39. As the grid density increases, the reattachment point progressively shifts downstream. The transition location of the separation bubble predicted using the "Medium" and "Fine" grids is closely aligned, indicating that further refinement will no longer influence the transition point. Consequently, the "Medium" grid was selected for subsequent investigations.

Figure 5.40 illustrates the grid topology of the computational domain and the mesh distribution in the vicinity of the trailing edge of the turbine blade. The total grid count amounts to 5.84 million. In the vicinity of the separation bubble, the maximum grid resolutions in the streamwise, normal, and spanwise directions satisfy the criteria of $\Delta x^+ < 15$, $\Delta y^+ < 15$, and $\Delta z^+ < 1$, respectively. For the unsteady simulation, the time step is set to 0.5 μs, corresponding to a non-dimensional time step of

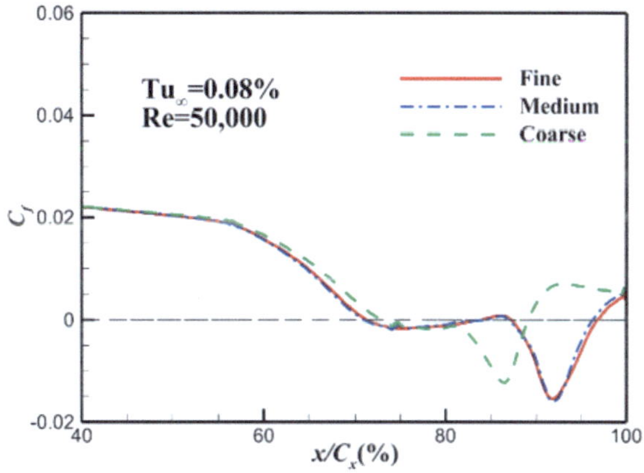

Fig. 5.39 Predicted skin friction distribution through different meshes

Table 5.6 Meshes considered in the grid sensitivity analysis

Mesh	Coarse	Medium	Fine
Growth rate in boundary layer	1.15	1.10	1.05
Number of streamwise nodes	361	541	681
Number of wall-normal nodes	49	81	113
Number of spanwise nodes	41	61	81

$\Delta t \cdot u_{\text{inf}}/c = 7.5 \times 10^{-6}$. The simulation is initially conducted using the r-transition model under the RANS framework. Subsequently, the γ − VLES model is applied to initiate unsteady computations. After the flow field has fully developed, time-averaged results are computed, with the time-averaging period set to approximately $15C_x/u_{\text{in}}$.

When a laminar boundary layer separates under unfavorable pressure gradient conditions, factors such as high shear stress and the absence of wall damping can cause a transition to occur in the form of free shear flow. The post-transition turbulence is then entrained back into the boundary layer, enhancing the mixing between laminar and turbulent flows, and ultimately leading to turbulent reattachment. This process—laminar separation, transition, and reattachment—characterizes the typical separation bubble transition mechanism. Separation bubbles are generally classified into two categories: short bubbles and long bubbles. Short bubbles have a localized effect on boundary layer flow, with the flow both upstream and downstream of the bubble closely resembling that of a non-separated case. Their primary function is to trigger transition, serving as an effective means to induce laminar-to-turbulent transition in the boundary layer, which can improve engine performance. In contrast, long bubbles typically form in proximity to the trailing edge of the blade, exhibit highly variable behavior, and exert a considerable influence on the global flow characteristics. This results in notable discrepancies in the surface pressure distribution across the entire blade in comparison to scenarios that do not involve separation bubbles. Furthermore, the presence of long bubbles results in considerable profile losses and

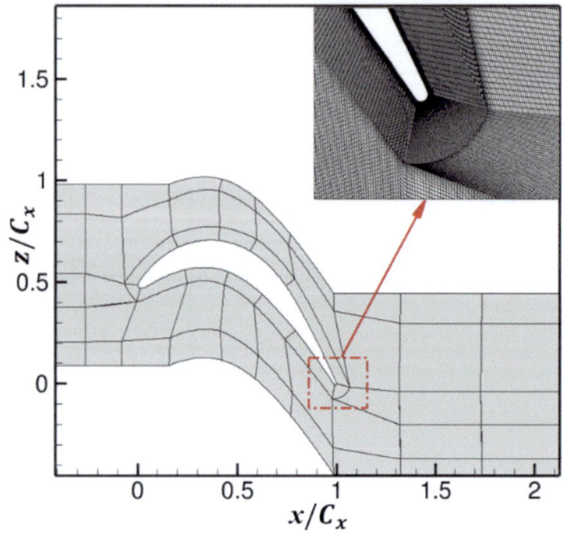

Fig. 5.40 Pak-B turbine blade computational domain size and local mesh distribution

a notable deviation from the desired exit flow angle, which ultimately impairs the overall performance of the system. In this instance, the separation bubble situated in close proximity to the trailing edge on the upper surface of the blade can be identified as a prototypical long bubble. The total pressure loss coefficient ξ is analyzed, defined as the difference between the total pressure at the inlet and outlet, non-dimensionalized by the outlet dynamic pressure: $\xi = \left(p_1^* - p_2^*\right)/\left(p_2^* - p_2\right)$, where p_1^*, p_2^*, and p_2 represent the inlet total pressure, outlet total pressure, and outlet static pressure, respectively. The $\gamma -$ VLES model predicts a total pressure loss of $\xi_{\gamma-\mathrm{VLES}} = 0.105$, while the r-transition model within the RANS framework predicts $\xi_\gamma = 0.075$. This result demonstrates that the $\gamma -$ VLES model more accurately represents the impact of long separation bubbles on the overall flow field, resulting in a greater predicted total pressure loss in comparison to the γ-transition model within the RANS framework.

A comparison of the time-averaged surface pressure distribution at the blade's mid-span, under the same inlet boundary conditions of turbulence intensity and Reynolds number, is presented in Fig. 5.41. The vertical axis represents the time-averaged pressure coefficient C_p, while the horizontal axis denotes $x/C_x(\%)$, where C_x is the blade chord length. The $\gamma -$ VLES model provides a highly accurate prediction of the separation bubble on the suction side, with the calculated $\overline{C_p}$ aligning closely with the experimental results. In contrast, the γ and $\gamma - \overline{Re_{\theta t}}$ transition models predict significantly smaller separation bubbles on the suction side and fail to accurately capture the strong unsteady characteristics of the trailing edge separation bubble. This demonstrates the $\gamma -$ VLES model's superior ability to replicate the complex flow structures present in the separation region.

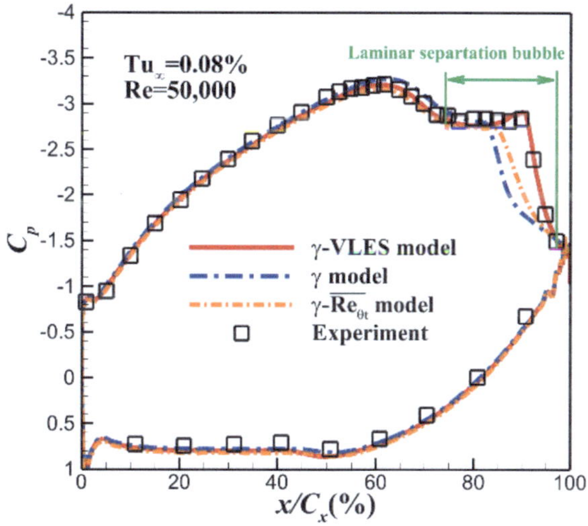

Fig. 5.41 Comparison of the averaged surface pressure coefficient at mid-span

A more detailed analysis of the results predicted by the $\gamma-$VLES model indicates that the separation bubble in the vicinity of the trailing edge on the suction side of the blade can be divided into two distinct regions, based on the point at which the time-averaged pressure begins to recover. In the initial region, preceding the onset of pressure recovery, the pressure coefficient C_p exhibits an almost plateau-like distribution with a slight upward trend. This indicates that a stronger reverse flow within the separation bubble occurs following the transition, which is a characteristic feature of long separation bubbles. The second region is situated within the recirculation zone of the time-averaged separation bubble, where an increase in surface static pressure is observed. The negative skin friction coefficient calculated on the "Medium" grid, shown in Fig. 5.39, provides a more direct representation of the flow separation and reattachment locations as predicted by the $\gamma-$ VLES model. These locations align closely with the length and development trend of the separation bubble as reflected in the surface pressure coefficient, thereby offering a consistent and comprehensive picture of the flow behavior.

Figure 5.42 illustrates a comparison of time-averaged velocity profiles at the mid-span of the blade, across seven different chordwise stations, computed using various transition models and benchmarked against experimental data from reference [35]. As the $\gamma-\overline{Re_{\theta t}}$ model demonstrated superior performance over the γ model in Fig. 5.41, the analysis focuses on velocity profiles predicted by the $\gamma-\overline{Re_{\theta t}}$ and $\gamma-$ VLES transition models. The horizontal axis represents the non-dimensional time-averaged velocity u/u_{in}, while the vertical axis denotes the non-dimensional wall-normal distance y_n/C_x, where y_n is the distance from the wall in the normal direction. The discussion focuses on the boundary layer velocity profiles along the suction side of the blade. Up to $x/C_x = 0.7$, the velocity profiles indicate a typical attached flow state, with results from both the $\gamma-\overline{Re_{\theta t}}$ and $\gamma-$ VLES transition models aligning well with experimental measurements. At stations $x/C_x = 0.75$

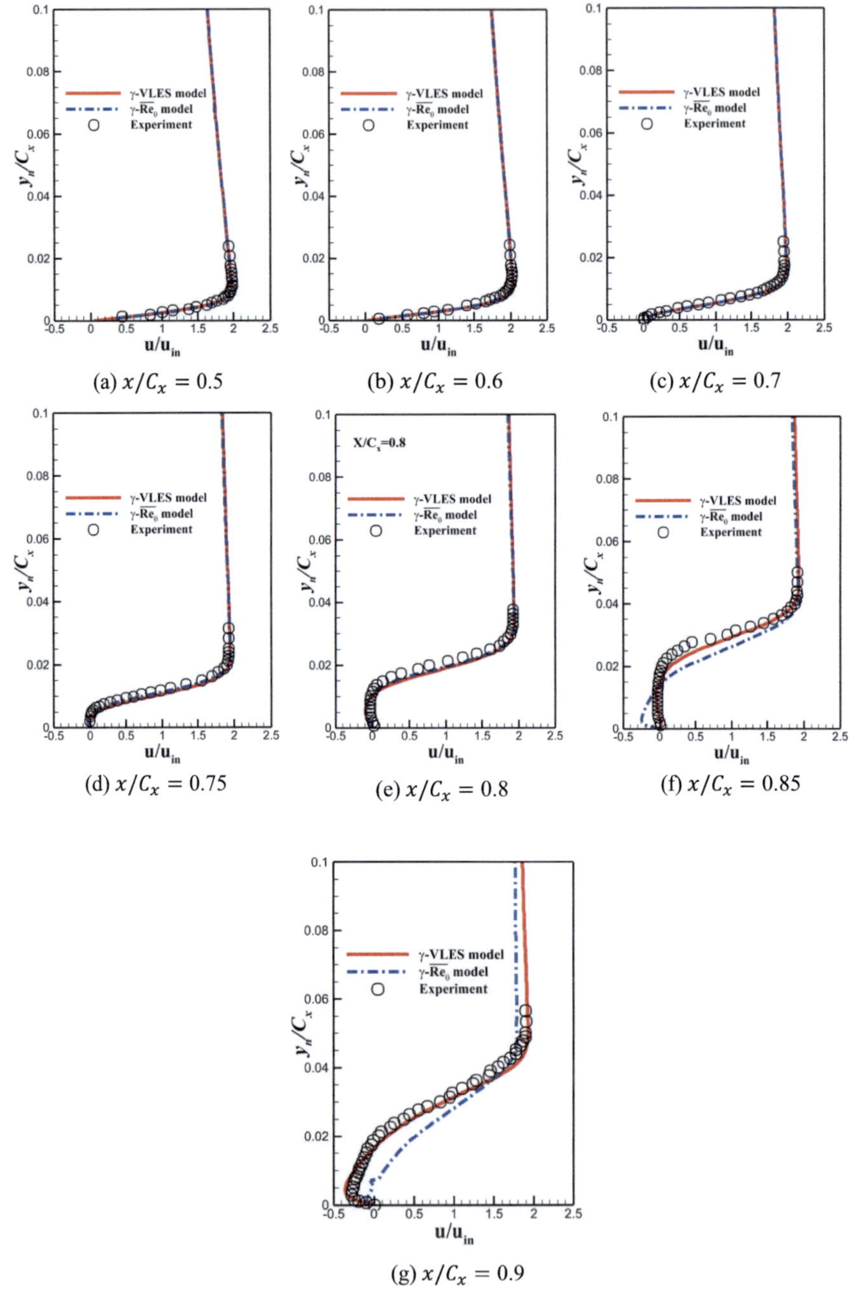

Fig. 5.42 Comparison of computed velocity profiles with experiment

and $x/C_x = 0.85$, the flow begins to separate, and the $\gamma - \overline{Re_{\theta t}}$ transition model continues to accurately pdict the boundary layer velocity profiles. At $x/C_x = 0.85.$, near-wall flow nearly stagnates, which corroborates the near-plateau behavior of the time-averaged pressure coefficient C_p. before its recovery, as seen in Fig. 5.41. At this position, the $\gamma -$ VLES model accurately predicts the flow behavior, while the $\gamma - \overline{Re_{\theta t}}$ model prematurely predicts reverse flow. At $x/C_x = 0.9$, ignificant reverse flow near the wall is observed, with the $\gamma -$ VLES model accurately capturing the boundary layer velocity distribution. In contrast, the $\gamma - \overline{Re_{\theta t}}$ model incorrectly predicts reattachment of the flow to the wall. This comparative analysis of the boundary layer velocity profiles underscores that the $\gamma -$ VLES model, specifically designed to handle large-scale flow separation, significantly enhances the accuracy of predicting flow separation and reattachment compared to the RANS-based $\gamma - \overline{Re_{\theta t}}$ transition model.

5.4 Summary

This chapter presents the development of an advanced $\gamma - A_r-$ VLES method, which extends the $\gamma - A_r$ transition model within a hybrid RANS/LES framework. The new method effectively integrates a roughness transition model with large eddy simulation (LES) techniques through several key innovations. These include the ability to control the solution approach within laminar boundary layers, to constrain the decay of free-stream turbulence, and to formulate transition influence functions specific to laminar separation bubbles. The $\gamma - A_r-$ VLES method has been developed with the specific aim of accurately simulating and analyzing the effects of transition and surface roughness on separated flow regimes. In order to guarantee the accurate prediction of transitions in attached flow scenarios, the method utilizes a coupling mechanism based on the intermittency factor γ, which restricts the $\gamma - A_r-$ VLES method to RANS solutions within the laminar boundary layer. In light of the fact that precise transition modeling is dependent upon the accurate representation of free-stream turbulence decay, a novel protection function is proposed. This function guarantees the use of the RANS model over the LES mode in free-stream regions, thereby facilitating the capture of turbulence decay with high fidelity. Moreover, to address the issue of premature VLES activation within laminar separation bubbles, which can result in delayed transition predictions, a switching function based on local turbulence intensity has been developed. This function refines the scale-switching mechanism of the VLES method, ensuring that VLES computations are appropriately confined to the laminar separation bubble region. The $\gamma - A_r-$ VLES method is validated through a series of numerical simulations, including smooth flat-plate attached flows, smooth flat-plate separated flows, flow around rough cylinders, flow around rough compressor blades, and flow around smooth turbine blades. A comprehensive analysis of the results indicates that the predictions generated by this hybrid method are in close alignment with experimental observations and LES data, thereby demonstrating both its adaptability and high degree of accuracy in simulating complex flow phenomena. Although the proposed hybrid method demonstrates robust predictive capabilities for flows over distributed rough surfaces, its application is currently constrained in the simulation of transition flows initiated by isolated roughness elements.

Chapter 6
Summary and Future Prospects

Abstract This chapter summarizes the key content of the entire book, distills its innovative contributions, and provides perspectives on future research directions.

6.1 Research Content and Conclusions

In flows around aircraft, wind turbines, and turbomachinery, the effects of laminar-to-turbulent boundary layer transition and surface roughness are commonly encountered. In specific operating conditions, the combination of transitional flow and large-scale boundary layer separation can result in complex three-dimensional unsteady phenomena, which have a significant impact on the aerodynamic performance, power generation, and thermal protection of these systems. To accurately predict the detailed characteristics of such complex flow fields while maintaining computational efficiency for engineering applications, this study introduces an advanced large eddy simulation (LES) method. This method is an effective means of capturing the influence of transition and surface roughness on the development of boundary layers and the evolution of separated flows. The $\gamma - A_r-$ VLES model, which has been developed, is capable of accurately predicting both bypass transition and transition induced by laminar separation bubbles. Furthermore, it effectively models the impact of surface roughness on the transition process. In instances where significant separation occurs on a large scale, the model is able to effectively capture the influence of transition and rough walls on the evolution of the flow, thereby providing a reliable tool for analyzing these complex flow conditions.

The principal findings of this study are presented below.

(1) This work examines the foundational principles behind hybrid RANS/LES methods, investigating the mathematical similarities between RANS and LES, and reviewing the key features of large eddy simulation (LES) construction. The shortcomings of the original VLES method, particularly its excessive reduction of Reynolds stress and its difficulty in properly transitioning to LES mode, are addressed. To resolve these issues, a scale control function based on turbulence integral length scale is developed to reduce the over-dissipation of Reynolds

© The Author(s), under exclusive license to Springer Nature Singapore Pte Ltd. 2025
J. Xu et al., *CFD-Compatible RANS/LES Modeling of Transitional and Separated Flows*, https://doi.org/10.1007/978-981-96-6886-1_6

stress. Additionally, model parameters are derived from local isotropic turbulence theory to ensure the correct activation of the LES mode. To improve accuracy in simulating boundary layers and separated flows, the model's filter scale is redefined, and a protective function is introduced. The efficacy of the novel VLES method was substantiated through the utilization of the Taylor–Green vortex (TGV) test cases at three distinct Reynolds numbers. The results demonstrate that the novel VLES method is capable of effectively capturing turbulent structures and their associated dissipation processes. Even at higher Reynolds numbers, the VLES method yields results that closely match DNS data, provided that appropriate time steps and sufficient grid resolution are used. Further validation of the VLES model's capability to simulate separated flows was conducted on additional cases, including the 30P30N multi-element airfoil, a single cylinder, and a square cylinder. The predicted surface pressure coefficients and velocity distributions in the slat cove for the 30P30N airfoil demonstrate an excellent degree of agreement with the corresponding experimental data. In the cylinder case, the predicted root mean square (RMS) of surface pressure coefficients exhibits a high degree of alignment with experimental measurements. The velocity distributions in the cylinder wake predicted by VLES also closely match experimental data and outperform the MDDES method, particularly in terms of separation zone length and spanwise vortex distribution. In the square cylinder case, the VLES method accurately predicts the velocity recovery in the wake, with predicted mean velocity, normal velocity, and RMS velocity fluctuations showing strong agreement with both experimental data and LES results. These findings substantiate the assertion that the recently devised VLES methodology represents a robust hybrid RANS/LES method, exhibiting considerable efficacy in the prediction of separated flows.

(2) A novel, concise, and effective three-equation transition model has been developed within the RANS framework. This model is capable of accurately capturing both bypass transition and separation bubble-induced transition phenomena, which are commonly observed in the flow fields of aircraft and turbomachinery. The model employs an intermittency factor transport equation to characterize the transition process. In order to derive new critical transition Reynolds numbers for bypass transition under high turbulence conditions, a comprehensive analysis of experimental data and DNS results was conducted. This analysis facilitated the establishment of a unified bypass transition criterion. For separation bubble-induced transition, a localized shape factor calculation function is introduced, integrating $H_{32} - R_{\delta_2}$ relationships to refine the model for laminar separation bubbles. This function enhances the calculation of the momentum thickness Reynolds number. To account for the effects of turbulence intensity and pressure gradients on separation bubbles, empirical correction functions are utilized, and model parameters are calibrated against experimental and DNS data. Additionally, adjustments to the source term in the turbulent kinetic energy transport equation are made to control the initiation and reattachment locations of transition within the separation bubble. The model was initially validated using attached flow flat plates, with mesh convergence studies conducted on the

T3A plate. These studies demonstrated that the results stabilized with sufficient mesh refinement, thereby ensuring accurate predictions. The model's capacity for predictive analysis in the context of bypass transition has been validated, while its ability to simulate T-S wave transition has been demonstrated through the presentation of results obtained from the T3A and S&K plates. Subsequent testing on pressure gradient-induced separation flat plates demonstrated that the model's predictions were in close alignment with experimental data, thereby validating its effectiveness in predicting separation bubble-induced transitions. The model has been validated extensively across a range of test cases, including T3L plates, V103 compressors, NACA0018 airfoils and SD-7003 airfoils. This confirms its capability to accurately characterize the size, transition location and reattachment processes of laminar separation bubbles under a range of turbulence intensities and flow Reynolds numbers.

(3) A comprehensive analysis of the influence of surface roughness on laminar-to-turbulent transition was conducted based on the transition model developed in this study. This analysis led to the development of the $\gamma - A_r$ roughness transition model. This model introduces a scalar transport equation that is linked to the equivalent roughness height, specifically the roughness growth factor A_r transport equation. By solving this equation, the distribution of A_r within the flow field is determined, allowing identification of the regions influenced by surface roughness. By applying boundary conditions at the rough wall surface, the distribution of A_r within the boundary layer can be obtained. The transition criteria are then modified using a functional relationship incorporating A_r, which captures the effect of roughness on the transition process. The functional relationship is calibrated using experimental data and data from direct numerical simulations (DNS). In light of the capacity of rough wall boundary conditions to represent transitional and fully rough surfaces, and their adaptability to laminar and transitional flows, the Knopp roughness model was selected for application within the transition model, specifically in the context of rough wall boundary conditions. Validation was initially conducted using rough flat plates with varying pressure gradients, wherein the predicted transition onset locations exhibited a high degree of concordance with experimental data. This confirmed the model's ability to accurately capture the influence of roughness height and pressure gradient on transition. For separation flat plates with varying roughness heights, the $\gamma - A_r$ transition model accurately predicted both the size of the separation bubble and the velocity profile, aligning well with experimental results, which demonstrates the model's effectiveness in capturing the influence of roughness on separation bubble-induced transitions. In the NREL-S814 case, the model successfully predicted the transition model change from laminar separation bubble-induced transition to bypass transition near the leading edge, due to the presence of the roughness band, with the predicted results closely matching experimental observations. This demonstrates the model's ability to predict the effects of rough leading edges on the pressure coefficient distribution on both the suction and pressure sides of the airfoil, as well as its capability to reasonably simulate the transition model changes caused by roughness. For

high- and low-pressure turbine blade flows, the model generally predicted transition onset locations in good agreement with experimental data. However, in some cases, the predicted transition locations were delayed, and the transition region was shorter than the experimental results. This discrepancy may stem from the model's limited consideration of roughness geometry density and streamline curvature. Nevertheless, the $\gamma - A_r$ roughness transition model demonstrates strong predictive capability for transition processes on rough surfaces across different Reynolds numbers, pressure gradients, and turbulence intensities, making it suitable for engineering applications such as rough surface wings and turbomachinery transitional flow simulations.

(4) Based on the VLES method introduced in this study, a new $\gamma - A_r -$ VLES model has been developed within the hybrid RANS/LES framework. This integrates the $\gamma - A_r$ roughness transition model to enable accurate simulation of the transition and the effects of surface roughness on separated flows. In this hybrid model, the scale control function of the VLES method may fail in local regions of the laminar boundary layer, necessitating the introduction of a coupling function based on the intermittency factor. This function constrains the model to operate in RANS model within the laminar boundary layer, preventing the activation of LES mode in those regions. In transition modeling, the correct capture of the decay of free-flow turbulence is critical for accurate predictions. To address this, a new safeguard has been introduced to ensure that the RANS model is used in the free flow region instead of the LES mode to properly account for turbulence decay. When laminar separation bubbles occur in the boundary layer, the local intermittency factor γ becomes 1.0, which activates the LES mode. Nevertheless, at this juncture, the RANS-based correction for separation-induced transition is ineffectual due to the absence of sufficient turbulence to precipitate transition, thereby yielding an inaccurate prediction of the transition process within the separation bubble. To address this limitation, a novel coupling function was introduced to guarantee that the RANS model remains active as the laminar boundary layer commences its separation. The LES mode is only enabled once the turbulence within the separation bubble has reached a sufficient intensity. In the context of validation cases, the model's predictions for smooth attached flow over a flat plate demonstrated complete alignment with the outcomes of the $\gamma - A_r$ transition model. This outcome substantiates the efficacy of the coupling function in the $\gamma - A_r -$ VLES model in maintaining the RANS-model computation within the laminar boundary layer and effectively constraining the free-stream turbulence decay. In simulations of separated flow over a rough flat plate, the VLES method yielded results that were markedly more closely aligned with LES data than those produced by the RANS method, thereby demonstrating superior agreement. In the case of rough cylinder flows, the presence of surface roughness gave rise to a bypass transition within the attached boundary layer, which in turn resulted in a modification of both the distribution of surface friction and the location of the wake separation point. The $\gamma - A_r -$ VLES model demonstrated superior alignment with experimental data compared to fully turbulent or RANS transition models,

and it accurately captured the cylinder's drag crisis phenomenon. In the case of the rough V103 compressor blade flow, an increase in the roughness height resulted in both the transition point and the turbulent reattachment point shifting upstream. The $\gamma - A_r-$ VLES model demonstrated efficacy in capturing these flow trends, with predictions for boundary layer development and separation flow exhibiting alignment with those of the LES. In the simulation of the flow over Pak-B turbine blades, the transition-separation hybrid model proposed in this book achieved excellent prediction results, with both the shape and location of the predicted separation bubble aligning consistently with experimental data.

6.2 Key Innovations

The main innovations of this study are as follows:

(1) An enhanced Very Large Eddy Simulation (VLES) model was developed based on the equilibrium turbulence assumption and local isotropic turbulence theory. By redefining the filter scale of the model and incorporating additional protection functions, it achieves improved accuracy in simulating both boundary layer and large-scale separated flows.
(2) New prediction criteria were developed for bypass transition in high turbulence environments and laminar separation bubble-induced transition. By integrating equivalent roughness theory with DNS and wind tunnel experimental data, a model was developed to capture the effects of distributed roughness elements on both bypass and bubble-induced transitions. This model showed good agreement with reference data in cases involving pressure gradient flat plates and compressor cascades.
(3) Integration of the roughness-transition model with the VLES method to develop a unified localized transition-separation flow prediction model suitable for rough surfaces. In typical test cases, such as the V103 compressor cascade, this model achieved simulation accuracy comparable to LES, but at a significantly reduced computational cost.

6.3 Future Research Directions

While this dissertation makes a substantial contribution to the field, there are still a number of issues that require further investigation. Further research could concentrate on the following areas:

(1) The present study employs primarily second-order upwind and central difference schemes for numerical calculations. To enhance the model's predictive precision for intricate flow scenarios, future research could investigate the integration of the hybrid RANS/LES approach with advanced numerical techniques to achieve heightened accuracy and resolution.

(2) In the context of turbine machinery flows, the substantial curvature of streamlines has the potential to exert a significant influence on both transition and separation phenomena. It would be beneficial for future research to incorporate the effects of streamline curvature into transition modeling, thus enhancing the accuracy and applicability of predictive models for complex flow scenarios.
(3) The established modeling framework must be validated to ensure its reliability in simulating flows around moving objects. Furthermore, its capability to model complex flows involving multiple interacting surfaces in relative motion must be enhanced.
(4) Advanced transition-separated flow prediction models can be employed for high-fidelity computations of aeroacoustic noise and combustion phenomena.

Appendix A
Boundary Layer Similarity Solutions

Since the transition criterion is the most important part in the process of boundary layer transition modeling, and the transition criterion needs to be established through non-local boundary layer variables such as boundary layer thickness, displacement thickness, momentum thickness and shape factors, which need to be obtained through boundary layer integration. In order to meet the massively parallel computing of modern CFD calculations, it is necessary to establish the relationship between local variables and non-local variables in the flow field, so as to realize the local solution of non-local variables. The boundary layer similarity solution provides a crucial way to solve this problem, so the boundary layer similarity solution must be accurately solved first [247].

Falkner-Skan

In instances where the boundary layer equations are amenable to similarity solutions, the velocity distribution within the boundary layer, $u(x, y)$, can be nondimensionalized through the utilization of appropriate scaling parameters. Following this process of nondimensionalization, the velocity profile $\frac{u(x,y)}{U(x)} = F(\eta)$ becomes identical at any position x, where $F(\eta)$ represents a universal function of the similarity variable η. This indicates that the velocity profile exhibits self-similar behavior across the boundary layer.

Similarity solutions are available for both the flat plate boundary layer and the flow in the vicinity of the stagnation point of a wedge. To illustrate, we may consider the boundary layer differential equation for steady, incompressible, two-dimensional flow over a flat plate, as presented in reference [248].

$$u\frac{\partial u}{\partial x} + v\frac{\partial v}{\partial y} = U\frac{dU}{dx} + \nu\frac{\partial^2 u}{\partial y^2} \tag{A.1}$$

© The Editor(s) (if applicable) and The Author(s), under exclusive license to Springer Nature Singapore Pte Ltd. 2025

J. Xu et al., *CFD-Compatible RANS/LES Modeling of Transitional and Separated Flows*, https://doi.org/10.1007/978-981-96-6886-1

$$\frac{\partial u}{\partial x} + \frac{\partial v}{\partial y} = 0 \tag{A.2}$$

Boundary conditions:
$y = 0; \quad u = v = 0$
$y = \infty; \; u = U$

In the equations above, x and y denote the directions along the flow over the flat plate and perpendicular to the plate surface, respectively. The variables u and v represent the velocity components in the streamwise and wall-normal directions, respectively, while $U = U(x)$ is the potential flow velocity at a given streamwise location x. To analyze the flow, we introduce a stream function $\varphi(x, y) = U(x)g(x)f(\eta)$ and a dimensionless wall-normal coordinate $\eta = \frac{y}{g(x)}$. Through derivation [249], it can be shown that the boundary layer equation possesses similarity solutions when $U = Cx^m$, where C is a constant and x represents the distance in the flow direction. In this case, the velocity profile function $f(\eta)$ satisfies the following governing relation:

$$f''' + ff'' + \beta_H\left(1 - f'^2\right) = 0 \tag{A.3}$$

The boundary conditions are:
$\eta = 0; \quad f(0) = f'(0) = 0$
$\eta = \infty; \; f'(\infty) = 1$

This equation is referred to as the Falkner-Skan (FS) equation, and its solutions are characterized by laminar, self-similar velocity profiles. Figure A.1 illustrates the solutions under various pressure gradients. The similarity variable is defined as $\eta = \frac{y}{g} = y\sqrt{\frac{m+1}{2}\frac{U}{vx}}$, where $m = \frac{\beta_H}{2\alpha - \beta_H}$, with β_H representing the pressure gradient parameter. For $\beta_H > 0$, the flow experiences a favorable pressure gradient; for $\beta_H = 0$, the pressure gradient is zero; and for $\beta_H < 0$, there is an adverse pressure gradient. The valid range for β_H is $-\, 0.19884 \le \beta_H \le 2.0$.

Falkner-Skan-Cooke

Building on the two-dimensional boundary layer similarity solution, J. C. Cooke extended it to three-dimensional boundary layers, leading to the development of the Falkner-Skan-Cooke (FSC) equation [250], which is now a fundamental concept in fluid mechanics. In this framework, the two-dimensional boundary layer equation describes the chordwise flow perpendicular to the leading edge, while Cooke's extended equation incorporates the spanwise flow along the leading edge. The detailed derivation of this equation is provided in reference [250], with the equation expressed as follows:

$$f''' + ff'' + \beta_H\left(1 - f'^2\right) = 0$$

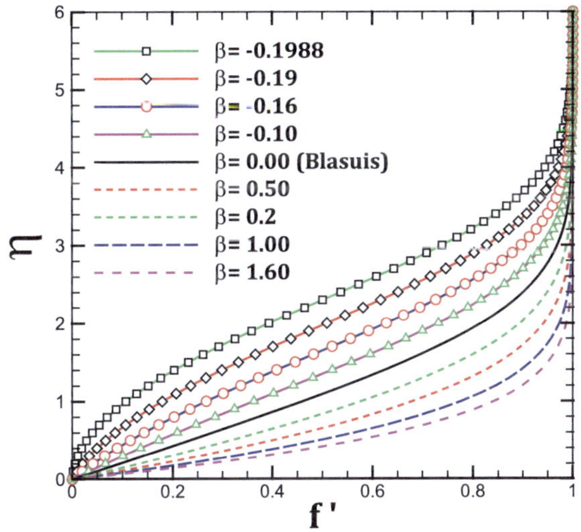

Fig. A.1 Similar velocity profile distributions of the FS laminar boundary layer under different pressure gradients

$$g'' + fg' = 0 \tag{A.4}$$

Boundary conditions:

$$f(0) = f'(0) = g(0) = 0, f'(\infty) = g(\infty) = 1$$

The FSC equation introduces a new spanwise velocity distribution function, $g(\eta) = \frac{W_s}{W_{s,\infty}}$, where W_s is the spanwise velocity and $W_{s,\infty}$ is the spanwise velocity at the outer edge of the boundary layer. As the value of the variable η approaches infinity, the value of W_s approaches $W_{s,\infty}$.

$$\beta_H = \frac{2m}{m+1}, \eta = y\sqrt{\frac{m+1}{2}\frac{U}{\nu x_c}} \tag{A.5}$$

The variable x_c represents the distance from the leading edge in the chordwise direction, which is perpendicular to the leading edge. The chordwise freestream velocity is defined as $U_{c,e} = C(x_c)^m$. The geometric sweep angle is denoted by φ_{sweep}, and the local sweep angle by φ; definitions of these directional and geometric parameters are provided for clarity. The boundary layer edge velocity, U_e, serves as the freestream velocity for the boundary layer and as the nondimensionalizing parameter. The chordwise and spanwise freestream velocities, corresponding to directions perpendicular and parallel to the leading edge, are given by $U_{c,e} = U_e\cos\varphi$ and $W_{s,e} = U_e\sin\varphi$, respectively. This setup defines the following velocity profile:

$$\frac{U_c}{U_{c,e}} = f'(\eta), \frac{W_s}{W_{s,e}} = g(\eta) \tag{A.6}$$

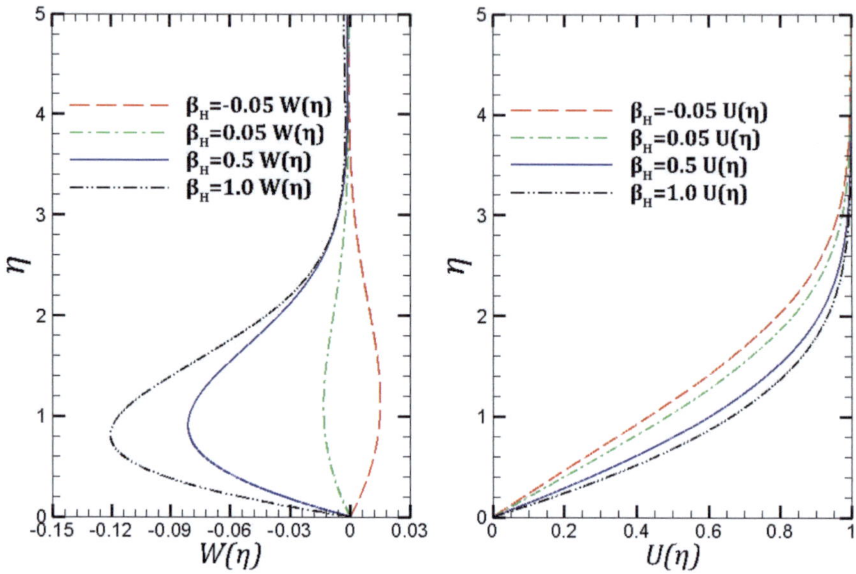

Fig. A.2 Solutions of the FSC equation under different pressure gradients for a 45° sweep angle

Thus, $U_c = f'(\eta)U_e\cos\varphi$ and $W_s = g(\eta)U_e\sin\varphi$, which gives the velocity components in the Cartesian coordinate system along the streamwise and spanwise directions as follows:

$$\begin{aligned} U &= U_c \cos\varphi + W_s \sin\varphi \\ &= U_e\left[f'(\eta)\cos^2\varphi + g(\eta)\sin^2\varphi\right] \end{aligned} \tag{A.7}$$

$$\begin{aligned} W &= -U_c \sin\varphi + W_s \cos\varphi \\ &= U_e\left[-f'(\eta) + g(\eta)\right]\sin\varphi\cos\varphi \end{aligned} \tag{A.8}$$

Thus, the nondimensional expression of the solution to the FSC equation in the Cartesian coordinate system is as follows:

$$U(\eta) = \tfrac{U}{U_e} = f'(\eta)\cos^2\varphi + g(\eta)\sin^2\varphi \tag{A.9}$$

$$W(\eta) = \tfrac{W}{U_e} = \left(-f'(\eta) + g(\eta)\right)\cos\varphi\sin\varphi \tag{A.10}$$

The FSC equation was solved for four distinct pressure gradients, resulting in the three-dimensional boundary layer similarity solutions for the FSC equation. The corresponding velocity profiles in the Cartesian coordinate system are illustrated in Fig. A.2.

Once the laminar boundary layer velocity profile has been established, it is essential to define the key characteristic Reynolds numbers for the boundary layer. Specifically, the Reynolds numbers based on the displacement thickness and momentum thickness of the boundary layer are defined as $\mathrm{Re}_{\delta 1} = \frac{U_e \delta_1}{\nu_e}$ and $\mathrm{Re}_{\theta} = \frac{U_e \theta}{\nu_e}$, respectively.

The velocity at the edge of the boundary layer, U_e, can be obtained based on the isentropic relation and the constant total enthalpy condition as defined in boundary layer theory:

$$\left(\frac{\gamma_H}{\gamma_H - 1}\right)\frac{P_\infty}{\rho_\infty} + \frac{U_\infty^2}{2} = \left(\frac{\gamma_H}{\gamma_H - 1}\right)\frac{P}{\rho_e} + \frac{U_e^2}{2} \tag{A.11}$$

$$\rho_e = \left[\frac{\rho_\infty^{\gamma_H}}{P_\infty} P\right]^{\frac{1}{\gamma_H}} \tag{A.12}$$

Here, γ_H represents the specific heat ratio, while P_∞, ρ_∞, and U_∞ denote the freestream pressure, density, and velocity, respectively.

References

1. SPALART PR, BOGUE D R. The Role of CFD in Aerodynamics, off-Design[J]. The Aeronautical Journal (1968), 2003, 107(1072): 323–329.
2. SAGAUT P, DECK S. Large Eddy Simulation for Aerodynamics: Status and Perspectives[J]. Philosophical Transactions of the Royal Society A: Mathematical, Physical and Engineering Sciences, 2009, 367(1899): 2849–2860.
3. FUJINO M. Design and Development of the HondaJet[J]. Journal of aircraft, 2005, 42(3): 755–764.
4. ZAKIT A, WISSINK J G, RODI W, et al. Direct Numerical Simulations of Transition in a Compressor Cascade: the Influence of Free-Stream Turbulence[J]. Journal of Fluid Mechanics, 2010, 665: 57–98.
5. RAMSAY R, HOFFMAN M, GREGOREK G. Effects of Grit Roughness and Pitch Oscillations on the S809 Airfoil[R]. [S.l.]: National Renewable Energy Lab. (NREL), Golden, CO (United States), 1995.
6. REUSS R, HOFFMAN M, GREGOREK G. Effects of Surface Roughness and Vortex Generators on the NACA4415 Airfoil[R]. [S.l.]: National Renewable Energy Lab. (NREL), Golden, CO (United States), 1995.
7. SPALART P R. Strategies for Turbulence Modelling and Simulations[J]. International Journal of Heat and Fluid Flow, 2000, 21(3): 252–263.
8. SCHUMANN U. Subgrid Scale Model for Finite Difference Simulations of Turbulent Flows in Plane Channels and Annuli[J]. Journal of Computational Physics, 1975, 18(4): 376–404.
9. SPEZIALE C G. Computing Non-Equilibrium Turbulent Flows with Time-Dependent RANS and VLES[C] // Fifteenth International Conference on Numerical Methods in Fluid Dynamics. 1996: 123–129.
10. SPEZIALE C G. Turbulence Modeling for Time-Dependent RANS and VLES: A Review[J]. AIAA Journal, 1998, 36(2): 173–184.
11. SPALART P R. Comments on the Feasibility of LES for Wings, and on a Hybrid RANS/LES Approach[C] //Proceedings of First AFOSR International Conference on DNS/LES. 1997.
12. CHAOUAT B. The State of the Art of Hybrid RANS/LES Modeling for the Simulation of Turbulent Flows[J]. Flow, Turbulence and Combustion, 2017, 99(2): 279–327.
13. FRÖHLICH J, VON TERZI D. Hybrid LES/RANS Methods for the Simulation of Turbulent Flows[J]. Progress in Aerospace Sciences, 2008, 44(5): 349–377.
14. SAGAUT P, TERRACOL M, DECK S. Multiscale and Multiresolution Approaches in Turbulence-LES, DES and Hybrid RANS/LES Methods: Applications and Guide-lines[M]. [S.l.]: World Scientific, 2013.

© The Editor(s) (if applicable) and The Author(s), under exclusive license to Springer Nature Singapore Pte Ltd. 2025

J. Xu et al., *CFD-Compatible RANS/LES Modeling of Transitional and Separated Flows*, https://doi.org/10.1007/978-981-96-6886-1

15. PIOMELLI U, BALARAS E, PASINATO H, et al. The Inner-Outer Layer Interface in Large-Eddy Simulations with Wall-Layer Models[J]. International Journal of Heat and Fluid Flow, 2003, 24(4): 538–550.
16. DAVIDSON L, DAHLSTRÖM S. Hybrid LES-RANS: An Approach to Make LES Applicable at High Reynolds Number[J]. International Journal of Computational Fluid Dynamics, 2005, 19(6): 415–427.
17. LARSSON J, LIEN F S, YEE E. The Artificial Buffer Layer and the Effects of Forcing in Hybrid LES/RANS[J]. International Journal of Heat and Fluid Flow, 2007, 28(6): 1443–1459.
18. SHUR M L, SPALART P R, STRELETS M K, et al. Synthetic turbulence generators for RANS-LES interfaces in zonal simulations of aerodynamic and aeroacoustic problems[J]. Flow, turbulence and combustion, 2014, 93(1): 63–92.
19. TEMMERMAN L, HADZIABDIC M, LESCHZINER M, et al. A Hybrid Two-Layer URANS-LES Approach for Large Eddy Simulation at High Reynolds Numbers[J]. International Journal of Heat and Fluid Flow, 2005, 26(2): 173–190.
20. ZHANG W, SHIH T I-P. Adaptive downstream Tensorial Eddy Viscosity for Hybrid Large-Eddy Simulations-Reynolds-Averaged-Navier-Stokes Simulations[J]. International Journal for Numerical Methods in Fluids, 2021, 93(6): 1825–1842.
21. ZHANG H, BACHMAN C, FASEL H. Application of a New Methodology for Simulations of Complex Turbulent Flows[C] //Fluids 2000 Conference and Exhibit. 2000: 2535.
22. FASEL H, SEIDEL J, WERNZ S. A Methodology for Simulations of Complex Turbulent Flows[J]. Journal of Fluids Engineering-ASME, 2002, 124(4): 933–942.
23. FASEL H F, von TERZI D A, SANDBERG R D. A Methodology for Simulating Compressible Turbulent Flows[J]. Journal of Applied Mechanics, 2005, 73 (3): 405–412.
24. ISRAEL D M. A New Approach for Turbulent Simulations in Complex Geometries[D]. [S.1.]: [s.n.], 2005.
25. HAN X, YE T, CHEN Y. Calibration of a New Very Large Eddy Simulation (VLES) Methodology for Turbulent Flow Simulation[J]. Science China, 2012, S5(10): 1905–1914.
26. HAN X, KRAJNOVI S. Validation of a Novel Very Large Eddy Simulation Method for Simulation of Turbulent Separated Flow[J]. International Journal for Numerical Methods in Fluids, 2013, 73(5): 436–461.
27. HAN X, KRAJNOVIć S. An Efficient Very Large Eddy Simulation Model for Simulation of Turbulent Flow[J]. International Journal for Numerical Methods in Fluids, 2013, 71(11): 1341–1360.
28. HAN X, KRAJNOVIć S. Very-Large-Eddy Simulation Based on Model[J]. AIAA Journal, 2015, 53(4): 1103–1108.
29. Zhang H D, Zhu J, Han S S. Research on Very-Large Eddy Simulation Technology for High Temperature Rise Combustor[J]. Aerospace Power, 2021
30. Zhou J, Wan P X, Han S S, et al. Very-Large Eddy Simulation of Turbulent Flow and Heat Transfer for Coupled Combustor-Turbine Components[J]. Journal of Propulsion Technology, 2022, 43(03): 246–256.
31. ZHANG K, JIN Y, HAN X, et al. Very-Large-Eddy Simulation of Nonreactive Turbulent Flow for Annular Trapped Vortex Combustor[J]. Journal of Applied Fluid Mechanics, 2022, 15(2):523–535.
32. LIU N-S, SHIH T-H. Turbulence Modeling for Very Large-Eddy Simulation[J]. AIAA Journal, 2006, 44(4): 687–697.
33. ZHANG Y, BAI J Q, HUA J, et al. Application of a continuous scale switch equation in steady and unsteady flow simulation[J]. ACTA Aerodynamica Sinica, 2014, 32(5):694–699.
34. SPALART P, ALLMARAS S. A One-Equation Turbulence Model for Aerodynamic Flows[C] // 30th Aerospace Sciences Meeting and Exhibit. 1992: 0439.
35. STRELETS M. Detached eddy Simulation of Massively Separated Flows[C] //39th Aerospace Sciences Meeting and Exhibit. 2001: 0879.
36. Li X X, Ren Y X. Detached-eddy Simulation $k - \varepsilon$ Turbulence Model and Its Application in Fire Simulations[J]. Tsinghua Univ (Sci & Tech), 2004, 44(08): 1126–1129.

37. MENTER F R, KUNTZ M, LANGTRY R. Ten Years of Industrial Experience with the SST Turbulence Model[J]. Turbulence, Heat and Mass Transfer, 2003, 4(1): 625–632.
38. SPALART P R, DECK S, SHUR M L, et al. A New Version of Detached-Eddy Simulation, Resistant to Ambiguous Grid Densities [J]. Theoretical and Computational Fluid Dynamics, 2006, 20(3): 181–195.
39. SHUR M L, SPALART P R, STRELETS M K, et al. A hybrid RANS-LES approach with delayed-DES and wall-modelled LES capabilities[J]. International Journal of Heat and Fluid Flow, 2008, 29(6): 1638–1649.
40. GRITSKEVICH MS, GARBARUK A V, SCHÜTZE J, et al. Development of DDES and IDDES formulations for the shear stress transport model[J]. Flow, turbulence and combustion, 2012, 88(3): 431–449.
41. YIN Z, REDDY K, DURBIN P A. On the Dynamic Computation of the Model Constant in Delayed Detached Eddy Simulation[J]. Physics of Fluids, 2015, 27(2): 025105.
42. YIN Z, GE X, DURBIN P. Adaptive Detached Eddy Simulation of Transition under the Influence of Free-Stream Turbulence and Pressure Gradient[J]. Journal of Fluid Mechanics, 2021, 915.
43. HE C, LIU Y, YAVUZKURT S. A Dynamic Delayed Detached-Eddy Simulation Model for Turbulent Flows[J]. Computers & Fluids, 2017, 146: 174–189.
44. WISSINK JG, RODI W. Direct Numerical Simulations of Transitional Flow in Turbomachinery [J]. Journal of Turbomachinery, 2006, 128(4): 668–678.
45. YUAN W, XU H, KHALID M, et al. A Parametric Study of LES on Laminar-Turbulent Transitional Flows Past an Airfoil[J]. International Journal of Computational Fluid Dynamics, 2006, 20(1): 45–54.
46. LARDEAU S, LESCHZINER M, ZAKI T. Large Eddy Simulation of Transitional Separated Flow over a Flat Plate and a Compressor Blade[J]. Flow, Turbulence and Combustion, 2012, 88(1-2): 19–44.
47. LIZ, JU Y, ZHANG C. Parallel Large Eddy Simulations of Transitional Flow in a Compressor Cascade with Endwalls[J]. Physics of Fluids, 2019, 31(11): 115104.
48. GERMANO M, PIOMELLI U, MOIN P, et al. A Dynamic Subgrid-Scale Eddy Viscosity Model[J]. Physics of Fluids A: Fluid Dynamics, 1991, 3(7): 1760–1765.
49. LILLY D K. A Proposed Modification of the Germano Subgrid-Scale Closure Method [J]. Physics of Fluids A: Fluid Dynamics, 1992, 4(3): 633–635.
50. MICHEL R. Etude de la Transition et Calcul de la Trainee des Profiles d'aile en Incompressible[J]. ONERA Publ, 1951, 58.
51. EPPLER R A, SOMERS D M. A Computer Program for the Design and Analysis of Low-Speed Airfoils: Vol 80210[M]. [S.1.]: National Aeronautics and Space Administration, Scientific and Technical Information Branch, 1980.
52. ABU-GHANNAM B, SHAW R. Natural Transition of Boundary Layers: The Effects of Turbulence, Pressure Gradient, and Flow History[J]. Journal of Mechanical Engineering Science, 1980, 22(5): 213–228.
53. SMITH A M O. Transition, Pressure Gradient and Stability Theory[J]. Douglas Aircraft Co., Report ES 26388, 1956.
54. VAN INGEN J. A Suggested Semi-Empirical Method for the Calculation of the Boundary Layer Transition Region[J]. Technische Hogeschool Delft, Vliegtuigbouwkunde, Report VTH-74, 1956.
55. ARNAL D, CASALIS G. Laminar-Turbulent Transition Prediction in Three-Dimensional Flows[J]. Progress in Aerospace Sciences, 2000, 36(2): 173–191.
56. CEBECI T. Stability and Transition: Theory and Application. Efficient Numerical Methods with Computer Programs[M]. [S.1.]: Heidelberg, Springer-Verlag Berlin, 2004.
57. BERTOLOTTIF P. Linear and Nonlinear Stability of Boundary Layers with Streamwise Varying Properties[D]. [S.1.]: [s.n.], 1990.
58. DRELA M, GILES M B. Viscous-Inviscid Analysis of Transonic and Low Reynolds Number Airfoils[J]. ATAA journal, 1987, 25(10): 1347–1355.

59. CODER JG, MAUGHMER M D. Computational Fluid Dynamics Compatible Transition Modeling Using an Amplification Factor Transport Equation[J]. AIAA Journal, 2014,52(11):2506–2512.
60. Zhang Z X, Dong Z N. Viscous Fluid Mechanics 3nd Ed.[M]. Beijing: Tsinghua University Press, 2011.
61. LIBBY P A. On the Prediction of Intermittent Turbulent Flows[J]. Journal of Fluid Mechanics, 1975, 68(2): 273–295
62. DOPAZO C. On Conditioned Averages for Intermittent Turbulent Flows[J]. Journal of Fluid Mechanics, 1977, 81(3): 433–438.
63. CHO J R, CHUNG M K. A K–ε–γ Equation Turbulence Model[J]. Journal of Fluid Mechanics, 1992, 237: 301–322.
64. STEELANT J, DICK E. Modeling of Laminar-Turbulent Transition for High Freestream Turbulence[J]. Journal of Fluids Engineering-ASME, 2000, 123(1): 22–30.
65. SUZEN Y B, XIONG G, HUANG P G. Predictions of Transitional Flows in Low-Pressure Turbines Using Intermittency Transport Equation[J]. AIAA Journal, 2002, 40(2): 254–266.
66. SUZEN Y B, HUANG P G, HULTGREN L S, et al. Predictions of Separated and Transitional Boundary Layers under Low-Pressure Turbine Airfoil Conditions Using an Intermittency Transport Equation[J]. Journal of Turbomachinery, 2003, 125(3): 455–464.
67. MENTER F R, LANGTRY R B, LIKKI S R, et al. A Correlation-Based Transition Model Using Local Variables - Part I: Model Formulation[J]. Journal of Turbomachinery, 2006, 128(3): 413–422.
68. LANGTRY R B, MENTER F R. Correlation-Based Transition Modeling for Unstructured Parallelized Computational Fluid Dynamics Codes[J]. AIAA Journal, 2009, 47(12): 2894–2906.
69. MENTER F R, SMIRNOV P E, LIU T, et al. A One-Equation Local Correlation-Based Transition Model[J]. Flow, Turbulence and Combustion, 2015, 95(4): 583–619
70. MAYLE R E. The Role of Laminar-Turbulent Transition in Gas Turbine Engines[J]. Journal of Turbomachinery, 1991, 113(4): 509–537.
71. MAYLE R E, SCHULZ A. The Path to Predicting Bypass Transition[J]. Journal of Turbomachinery, 1997, 119(3): 405–411.
72. MALKIEL E, MAYLER. Transition in a Separation Bubble[J]. Journal of Turbomachinery, 1996, 118: 752–759.
73. WALTERS D K, LEYLEK JH. A New model for Boundary Layer Transition Using a Single-Point RANS Approach[J]. Journal of Turbomachinery, 2004, 126(1): 193–202
74. WALTERS D K, COKLJAT D. A Three-Equation Eddy-Viscosity Model for Reynolds-Averaged Navier-Stokes Simulations of Transitional Flow[J]. Journal of Fluids Engineering-ASME, 2008, 130(12): 12140101–12140114.
75. WANG L, FU S. Development of an Intermittency Equation for the Modeling of the Supersonic/Hypersonic Boundary Layer Flow Transition[J]. Flow, Turbulence and Combustion, 2011, 87(1): 165–187.
76. WANG L, FU S, CARNARIUS A, et al. A Modular RANS Approach for Modelling Laminar-Turbulent Transition in Turbomachinery Flows[J]. International Journal of Heat and Fluid Flow, 2012, 34: 62–69.
77. FU S, WANG L. RANS Modeling of High-Speed Aerodynamic Flow Transition with Consideration of Stability Theory[J]. Progress in Aerospace Sciences, 2013, 58: 36–59.
78. JUNTASARO E, NGIAMSOONGNIRN K. A New Physics-Based $\gamma - k_L$ Transition Model[J]. International Journal of Computational Fluid Dynamics, 2014, 28(5): 204–218
79. XU J, BAI J, QIAO L, et al. Fully Local Formulation of a Transition Closure Model for Transitional Flow Simulations[J]. AIAA Journal, 2016, 54(10): 3015–3023.
80. XU J, BAI J, QIAO L, et al. Development of a Computational Fluid Dynamics Compatible Mathematical Model for Boundary Layer Transitional Flows in Low-Disturbance Environment[J]. Aerospace Science and Technology, 2019, 86: 487–496.
81. JACOBS R G, DURBIN P A. Simulations of Bypass Transition[J]. Journal of Fluid Mechanics, 2001, 428: 185–212.

82. ZAKITA, DURBIN P A. Mode Interaction and the Bypass Route to Transition[J]. Journal of Fluid Mechanics, 2005, 531: 85–111.
83. DURBIN P. An Intermittency Model for Bypass Transition[J]. International Journal of Heat and Fluid Flow, 2012, 36: 1–6.
84. GE X, AROLLA S, DURBIN P. A Bypass Transition Model Based on the Intermittency Function[J]. Flow, Turbulence and Combustion, 2014, 93(1): 37–61.
85. KUBACKI S, DICK E. An Algebraic Model for Bypass Transition in Turbomachinery Boundary Layer Flows[J]. International Journal of Heat and Fluid Flow, 2016, 58: 68–83.
86. KUBACKI S, DICK E. An Algebraic Intermittency Model for Bypass, Separation-Induced and Wake-Induced Transition[J]. International Journal of Heat and Fluid Flow, 2016,62: 344–361.
87. WILCOX D C. Turbulence and Transition Modeling for High-Speed Flows: ADA266748[R]. [S.1.]: DCW Industries Inc. La Canada CA, 1993.
88. WILCOX D C, OTHERS. Turbulence Modeling for CFD: Vol 2nd[M]. [S.1.]: DCW industries La Canada, CA, 1998.
89. NIKURADSE J. Laws of Flows in Rough Pipes: Technical Memorandum 1292[R]. [S.1.]: NACA, 1937.
90. BONS J P, TAYLOR R P, MCCLAIN S T, et al. The Many Faces of Turbine Surface Roughness[J]. ASME Journal of Turbomachinery, 2001, 123(4): 739–748.
91. LU M-H, LIOU W W. New Two-Equation Closure for Rough-Wall Turbulent Flows Using the Brinkman Equation[J]. AIAA Journal, 2009, 47(2): 386–398.
92. BUSSE A, SANDHAM N D. Parametric Forcing Approach to Rough-Wall Turbulent Channel Flow[J]. Journal of Fluid Mechanics, 2012, 712: 169–202.
93. BONS J, MCCLAIN S, WANG Z, et al. A Comparison of Approximate Versus Exact Geometrical Representations of Roughness for CFD Calculations of C_f and St[J]. Journal of Turbomachinery, 2008, 130: 021024–1.
94. SCHLICHTING H. Experimentelle untersuchungen zum rauhigkeitsproblem[J]. Ingenieur-Archiv, 1936, 7(1): 1–34.
95. ROBERSON J A. Surface Resistance as a Function of the Concentration and Size of Roughness Elements[D]. [S.1.]: [s.n.], 1961.
96. FINSON M. A Model for Rough Wall Turbulent Heating and Skin Friction[C] // 20th Aerospace Sciences Meeting. 1982: 1982–0199.
97. COLEMAN H W, HODGE B, TAYLOR R P. Generalized Roughness Effects on Turbulent Boundary Layer Heat Transfer. A Discrete Element Predictive Approach for Turbulent Flow over Rough Surfaces[R]. [S.l.]: Mississippi State University Report MSSU-TFD-84–1, 1983.
98. MCCLAIN S T, HODGE B K, BONS J P. Predicting Skin Friction and Heat Transfer for Turbulent Flow Over Real Gas Turbine Surface Roughness Using the Discrete Element Method[J]. Journal of Turbomachinery, 2004, 126(2): 259–267.
99. MCCLAIN S T, COLLINS S P, HODGE B K, et al. The Importance of the Mean Elevation in Predicting Skin Friction for Flow Over Closely Packed Surface Roughness[J]. Journal of Fluids Engineering, 2005, 128(3): 579–586.
100. DURBIN P A, MEDIC G, SEO J-M, et al. Rough Wall Modification of Two-Layer [J] Journal of Fluids Engineering-ASME, 2001, 123(1): 16–21.
101. AUPOIX B, SPALART P R. Extensions of the Spalart-Allmaras Turbulence Model to Account for Wall Roughness[J]. International Journal of Heat and Fluid Flow, 2003, 24(4): 454–462.
102. HELLSTEN A, LAINE S. Extension of $k-\omega$ Shear-Stress Transport Turbulence Model for Rough-Wall Flows[J]. AIAA Journal, 1998, 36(9): 1728–1729.
103. KNOPP T, EISFELD B, CALVO J B. A New Extension for k-w Turbulence Models to Account for Wall Roughness[J]. International Journal of Heat and Fluid Flow, 2009, 30(1): 54–65.
104. AUPOIX B. Roughness Corrections for the $k-\omega$ Shear Stress Transport Model: Status and Proposals[J]. Journal of Fluids Engineering, 2015, 137(2): 021202–10.
105. COLEBROOK C F, WHITE C M, TAYLOR G I. Experiments with Fluid Friction in Roughened Pipes [J]. Proceedings of the Royal Society of London. Series A - Mathematical and Physical Sciences, 1937, 161(906): 367–381.

106. BOYLER J, STRIPF M. Simplified Approach to Predicting Rough Surface Transition[J]. Journal of Turbomachinery, 2009, 131(4): 041020.
107. STRIPF M, SCHULZ A, BAUER H J, et al. Extended Models for Transitional Rough Wall Boundary Layers With Heat Transfer-Part I: Model Formulations[J]. Journal of Turbomachinery, 2009, 131(3): 031016.
108. GE X, DURBIN P A. An Intermittency Model for Predicting Roughness Induced Transition[J]. International Journal of Heat and Fluid Flow, 2015, 54: 55–64.
109. WEIL, GE X, GEORGE J, et al. Modeling of Laminar-Turbulent Transition in Boundary Layers and Rough Turbine Blades[J]. Journal of Turbomachinery, 2017, 139(11): 111009.
110. DASSLER P, KOZULOVIC D, FIALA A. Transport Equation for Roughness Effects on Laminar-Turbulent Transition[C] // Fifth European Conference on Computational Fluid Dynamics. 2010, pp: 323–330.
111. DASSLER P, KOZULOVIC D, FIALA A. An Approach for Modelling the Roughness-Induced Boundary Layer Transition Using Transport Equations[C] // Euro. Congress on Comp. Methods in Appl. Sciences and Engineering, ECCOMAS. 2012.
112. LANGEL C M, CHOW R, VAN DAM C C P. Further Developments to a Local Correlation Based Roughness Model for Boundary Layer Transition Prediction[C] // 53rd AIAA Aerospace Sciences Meeting. 2015: 1232.
113. LANGEL C M, CHOW R, van DAM C P, et al. RANS Based Methodology for Predicting the Influence of Leading Edge Erosion on Airfoil Performance: SAND-2017–11289[R]. [S.1.]: Sandia National Lab.(SNL-NM), Albuquerque, NM (United States), 2017.
114. YANG M, XIAO Z. Distributed Roughness Induced Transition on Wind-Turbine Airfoils Simulated by Four-Equation $k - \omega - \gamma - A_r$ Transition Model[J]. Renewable Energy, 2019, 135:1166–1177.
115. WANG L, FU S. Modelling Flow Transition in a Hypersonic Boundary Layer with Reynolds-Averaged Navier-Stokes Approach[J]. Science in China Series G: Physics, Mechanics and Astronomy, 2009, 52(5): 768–774.
116. LIU Z, ZHAO Y, CHEN S, et al. Predicting Distributed Roughness Induced Transition with a Four-equation Laminar Kinetic Energy Transition Model[J]. Aerospace Science and Technology, 2020, 99: 105736.
117. FASEL H F, von TERZI D A, SANDBERG R D. A Methodology for Simulating Compressible Turbulent Flows[J]. Journal of Applied Mechanics and Technical Physics, 2006, 73: 405–412.
118. GATSKI T B, SPEZIALE C G. On Explicit Algebraic Stress Models for Complex Turbulent Flows[J]. Journal of Fluid Mechanics, 1993, 254: 59–78.
119. ISRAEL D M. A New Approach for Turbulent Simulations in Complex Geometries[D]. [S.1.]: The University of Arizona, 2005.
120. GATSKIT, JONGEN T. Nonlinear Eddy Viscosity and Algebraic Stress Models for Solving Complex Turbulent Flows[J]. Progress in Aerospace Sciences, 2000, 36(8): 655–682.
121. HSIEH K-J. A Hybrid Numerical Simulation Approach for Turbulent Flows over Building-Like Obstacles[D]. [S.l.]: the University of Waterloo, 2008.
122. HSIEH K-J, LIEN F-S, YEE E. Towards a Unified Turbulence Simulation Approach for Wall-Bounded Flows[J]. Flow, Turbulence and Combustion, 2010, 84(2): 193–218.
123. BATTEN P, GOLDBERG U, CHAKRAVARTHY S. Interfacing Statistical Turbulence Closures with Large-Eddy Simulation[J]. AIAA Journal, 2004, 42(3): 485–492.
124. ISRAEL D, FASEL H. Numerical investigation of turbulent separation control using periodic disturbances[C] //40th AIAA Aerospace Sciences Meeting & Exhibit. 2002: 409.
125. MAGNIENT J-C, SAGAUT P, DEVILLE M. A Study of Built-in Filter for Some Eddy Viscosity Models in Large-Eddy Simulation[J]. Physics of Fluids, 2001, 13(5): 1440–1449.
126. Zhang Z S, Cui G X, Xu C X. Theory and Applications of Large Eddy Simulation. Beijing: Tsinghua University Press (in Chinese), 2008.
127. BREUER M, JOVIČIĆ N, MAZAEV K. Comparison of DES, RANS and LES for the separated flow around a flat plate at high incidence[J]. International Journal for Numerical Methods in Fluids, 2003, 41(4): 357–388.

128. REDDY KR, RYONJA, DURBINPA. A DDES Model with a Smagorinsky-Type Eddy Viscosity Formulation and Log-Layer Mismatch Correction[J]. International Journal of Heat and Fluid Flow, 2014, 50: 103–113.
129. MENTER F, KUNTZ M. Adaptation of Eddy-Viscosity Turbulence Models to Unsteady Separated Flow behind Vehicles[G] //The Aerodynamics of Heavy Vehicles: Trucks, Buses, and Trains. [S.1.]: Springer, 2004: 339–352.
130. MENTER F R. Two-Equation Eddy-Viscosity Turbulence Models for Engineering Applications[J]. AIAA Journal, 1994, 32(8): 1598 1605.
131. ROE P L. Approximate Riemann Solvers, Parameter Vectors, and Difference Schemes[J]. Journal of Computational Physics, 1981, 43(2): 357–372
132. VAN LEER B. Towards the Ultimate Conservative Difference Scheme. V. A Second-Order Sequel to Godunov's Method[J]. Journal of Computational Physics, 1979, 32(1): 101–136.
133. BEAM R M, WARMING R. An Implicit Factored Scheme for the Compressible Navier-Stokes Equations[J]. AIAA Journal, 1978,16(4): 393–402.
134. PULLIAM T H, CHAUSSEE D S. A Diagonal Form of an Implicit Approximate-Factorization Algorithm[J]. Journal of Computational Physics,1981, 39(2): 347–363.
135. JAMESON A. Time Dependent Calculations Using Multigrid, with Applications to Unsteady Flows past Airfoils and Wings[C] // 10th Computational Fluid Dynamics Conference. 1991: 1596.
136. Zhang Y. A New Turbulence Model for External Flow Simulation of Aircraft[D]. [S.1.]: Northwestern Polytechnical University Press (in Chinese), 2014.
137. Xu J K. Modelling Methods and Application Research of Boundary Layer Transition in Multi-Speed Range based on RANS Equations[D]. [S.1.]: Northwestern Polytechnical University, 2017.
138. TAYLOR G I, GREEN A E. Mechanism of the Production of Small Eddies from Large Ones[J]. Proceedings of the Royal Society of London. Series A, Mathematical and Physical Sciences, 1937, 158(895): 499–521.
139. BRACHET ME, MEIRON D I, ORSZAG S A, et al. Small-Scale Structure of the Taylor-Green Vortex[J]. Journal Fluid Mechanics, 1983, 130: 411–452.
140. BRACHET ME. Direct simulation of three-dimensional turbulence in the Taylor Green vortex[J] Fluid Dynamics Research, 1991, 8(1): 1–8.
141. DRIKAKIS D, FUREBY C, GRINSTEIN F F, et al. Simulation of Transition and Turbulence Decay in the Taylor-Green Vortex[J]. Journal of Turbulence, 2007, 8: N20.
142. WANG Z J, FIDKOWSKI K, ABGRALL R, et al. High-Order CFD Methods: Current Status and Perspective[J]. International Journal for Numerical Methods in Fluids, 2013, 72(8): 811–845.
143. Liu T X, Ma B F. Suitability of Low-order Numerical Schemes in Large Eddy Simulations[J]. Chinese Journal of Computational Physics, 2014(3): 307–313.
144. BULL JR, JAMESON A. Simulation of the Taylor-Green Vortex Using High-Order Flux Reconstruction Schemes[J]. AIAA Journal, 2015, 53(9): 2750–2761.
145. HUNT JC, WRAY A A, MOIN P. Eddies, Streams, and Convergence Zones in Turbulent Flows[J]. Studying Turbulence Using Numerical Simulation Databases, 2. Proceedings of the 1988 Summer Program, 1988.
146. VALAREZO W, DOMINIK C, MCGHEE R, et al. Multi-Element Airfoil Optimization for Maximum Lift at High Reynolds Numbers[C] // 9th Applied Aerodynamics Conference. 1991: 3332.
147. CHOUDHARIM, VISBAL M. Second Workshop on Benchmark Problems for Airframe Noise Computations[J]. Online. URL https://info.aiaa.org/tac/ASG/FDTC/DG/BE-CAN files_/BANCIl.Htm, 2011.
148. MURAYAMA M, NAKAKITA K, YAMAMOTO K, et al. Experimental Study on Slat Noise from 30P30N Three-Element High-Lift Airfoil at JAXA Hard-Wall Lowspeed Wind Tunnel[C] //20th AIAA/CEAS Aeroacoustics Conference. 2014: 2080.
149. PASCIONI K, CATTAFESTA L N, CHOUDHARI M M. An Experimental Investigation of the 30P30N Multi-Element High-Lift Airfoil[C] // 20th AIAA/CEAS Aeroacoustics Conference. 2014: 3062.

150. PASCIONI K A, CATTAFESTA L N. Aeroacoustic Measurements of Leading-Edge Slat Noise[C] // 22nd AIAA/CEAS Aeroacoustics Conference. 2016: 2960.
151. TERRACOL M, MANOHA E, MURAYAMA M, et al. Aeroacoustic Calculations of the 30p30n High-Lift Airfoil Using Hybrid RANS/LES Methods: Modeling and Grid Resolution Effects[C] //21st AIAA/CEAS Aeroacoustics Conference. 2015: 3132.
152. JENKINS L, KHORRAMI M, CHOUDHARI M, et al. Characterization of Unsteady Flow Structures around Tandem Cylinders for Component Interaction Studies in Airframe Noise[C] // 11th AIAA/CEAS Aeroacoustics Conference. 2005: 2812.
153. VATSA V, LOCKARD D. Assessment of Hybrid RANS/LES Turbulence Models for Aeroacoustics Applications[C] // 16th AIAA/CEAS Aeroacoustics Conference. 2010: 4001.
154. DAVIS R W, MOORE E F. A Numerical Study of Vortex Shedding from Rectangular Cylinders[J]. Journal of Fluid Mechanics, 1982, 116: 475–506.
155. LYN D A, RODI W. The Flapping Shear Layer Formed by Flow Separation from the Forward Corner of a Square Cylinder[J]. Journal of Fluid Mechanics, 1994, 267: 353–376.
156. OKAJIMA A. Strouhal Numbers of Rectangular Cylinders [J. Journal of Fluid Mechanics, 1982, 123: 379 – 398.
157. DAVIS R W, MOORE E F, PURTELL L P. A Numerical-Experimental Study of Confined Flow around Rectangular Cylinders [J]. The Physics of Fluids, 1984, 27(1): 46–59.
158. NAKAGAWA S, NITTA K, SENDA M. An Experimental Study on Unsteady Turbulent near Wake of a Rectangular Cylinder in Channel Flow[J]. Experiments in Fluids, 1999, 27(3): 284–294.
159. KIM D-H, YANG K-S, SENDA M. Large Eddy Simulation of Turbulent Flow Past a Square Cylinder Confined in a Channel[J]. Computers & fluids, 2004, 33(1): 81–96
160. XU J, BAI J, FU Z, et al. Parallel Compatible Transition Closure Model for High-Speed Transitional Flow[J]. AIAA Journal, 2017, 55(9): 3040–3050.
161. CHANG M, QIAO L, XU J, et al. A Fully-Local Transition Prediction Model for High-Turbulence Disturbance Environment[C] // Asia-Pacific International Symposium on Aerospace Technology. 2018: 682–702.
162. XU J, QIAO L, BAI J. Improved Local Amplification Factor Transport Equation for Stationary Crossflow Instability in Subsonic and Transonic Flows[J]. Chinese Journal of Aeronautics, 2020, 33(12): 3073–3081.
163. QIAOL, XU J, BAI J, et al. Fully Local Transition Closure Model for Hypersonic Boundary Layers Considering Crossflow Effects[J]. AIAA Journal, 2021, 59(5): 1692–1706.
164. CODER J, MAUGHMER M. One-Equation Transition Closure for Eddy-Viscosity Turbulence Models in CFD[C] //50th AIAA Aerospace Sciences Meeting. 2012: AIAA 2012–0672.
165. SIMONI D, UBALDI M, ZUNINO P, et al. Free-Stream Turbulence Effects on the Boundary Layer of a High-Lift Low-Pressure-Turbine Blade[J]. Journal of Thermal Science, 2016, 25(3): 195–206.
166. SIMONI D, LENGANI D, UBALDI M, et al. Inspection of the Dynamic Properties of Laminar Separation Bubbles: Free-Stream Turbulence Intensity Effects for Different Reynolds Numbers[J]. Experiments in Fluids, 2017, 58(6): 1–14.
167. ISTVAN MS, KURELEK JW, YARUSEVYCH S. Turbulence Intensity Effects on Laminar Separation Bubbles Formed over an Airfoil[J]. AIAA Journal, 2018, 56(4): 1335–1347.
168. MCAULIFFE B R, YARAS M I. Transition Mechanisms in Separation Bubbles Under Low- and Elevated-Freestream Turbulence[J]. Journal of Turbomachinery, 2010, 132(1): 011004.
169. BALZER W, FASEL H F. Numerical Investigation of the Role of Free-Stream Turbulence in Boundary-Layer Separation[J]. Journal of Fluid Mechanics, 2016, 801: 289–321.
170. SENGUPTA A, TUCKER P. Effects of Forced Frequency Oscillations and Free Stream Turbulence on the Separation-Induced Transition in Pressure Gradient Dominated Flows[J]. Physics of Fluids, 2020, 32(10): 104105.
171. HARTREE D R. On an Equation Occurring in Falkner and Skan's Approximate Treatment of the Equations of the Boundary Layer[J]. Math Proc. Cambridge, 1937, 33(2): 223–239.
172. CODER J G. Enhancement of the Amplification Factor Transport Transition Modeling Framework[C] // 55th AIAA Aerospace Sciences Meeting. 2017: AIAA 2017–1709.

173. EPPLER R. Turbulent Airfoils for General Aviation[J]. Journal of Aircraft, 1978, 15(2): 3–99.
174. DINI P. A Computationally Efficient Modelling of Laminar Separation Bubbles[M]. [S.1.]: The Pennsylvania State University, 1990.
175. SCHUBAUER G B, KLEBANOFF PS. Contributions on the Mechanics of Boundary-Layer Transition: NACA TN-3489[R]. 1955.
176. SAVILL A. Some Recent Progress in the Turbulence Modelling of By-pass Transition [J]. Near-Wall Turbulent Flows. Edited by R. M. C. So, C. G. Speziale, and B. E. Launder, 1993: 829–848.
177. SAVILL A. One-Point Closures Applied to Transition, Turbulence and Transition Modeling[J]. Edited by M. Hallbäck et al., 1996: 233–268.
178. LOU W, HOURMOUZIADIS J. Separation Bubbles Under Steady and Periodic-Unsteady Main Flow Conditions[J]. Journal of Turbomachinery, 2000, 122(4): 634–643.
179. VOLINO R J, HULTGREN L S. Measurements in Separated and Transitional Boundary Layers Under Low-Pressure Turbine Airfoil Conditions[J]. Journal of Turbomachinery, 2000, 123(2): 189–197.
180. SAVILL A. Further Progress in the Turbulence Modelling of By-pass Transition[G] // Engineering Turbulence Modelling Experiments. [S.1.]: Elsevier, 1993: 583–592.
181. VICEDO J, VILMIN S, DAWES W N, et al. Intermittency Transport Modeling of Separated Flow Transition[J]. Journal of Turbomachinery, 2004, 126(3): 424–431.
182. DORNEY D J, LAKE JP, KING P I, et al. Experimental and Numerical Investigation of Losses in Low-Pressure Turbine Blade Rows[J]. Interntional Journal of Turbo and Jet-Engines, 2000, 17(4): 241–254.
183. HUANG J, CORKE T C, THOMAS F O. Plasma Actuators for Separation Control Low-Pressure Turbine Blades[J]. AIAA Journal, 2006, 44(1): 51–57.
184. HILGENFELD L, PFITZNER M. Unsteady Boundary Layer Development due to Wake Passing Effects on a Highly Loaded Linear Compressor Cascade[J]. Journal of Turbomachinery, 2004, 126(4): 493–500.
185. ISTVAN MS, YARUSEVYCH S. Effects of Free-Stream Turbulence Intensity on Transition in a Laminar Separation on Bubble Formed over an Airfoil[J]. Experiments in fluids, 2018, 59(3): 1–21.
186. TANGERMANN E, KLEIN M. Numerical Simulation of Laminar Separation on a NACA0018 Airfoil in Freestream Turbulence[C] // AIAA Scitech 2020 Forum. 2020: 2064.
187. OL M, MCCAULIFFE B, HANFF E, et al. Comparison of Laminar Separation Bubble Measurements on a Low Reynolds Number Airfoil in Three Facilities[C] // 35th ATAA Fluid Dynamics Conference and Exhibit. 2005: 5149
188. BURGMANN S, DANNEMANN J, SCHRÖDER W. Time-Resolved and Volumetric PIV Measurements of a Transitional Separation Bubble on an SD7003 Airfoil[J]. Experiments in Fluids, 2008, 44(4): 609–622.
189. HAIN R, KÄHLER C, RADESPIEL R. Dynamics of Laminar Separation Bubbles at Low-Reynolds-Number Aerofoils[J]. Journal of Fluid Mechanics, 2009, 630: 129–153.
190. ZILLI J, SUTTON D M, LAVOIE P. Effect of Freestream Turbulence on Laminar Separation Bubbles and Flow Transition on an SD7003 Airfoil at Low Reynolds Numbers[C] //55th ATAA Aerospace Science Meeting. 2017: 0302.
191. CATALANO P, TOGNACCINI R. Influence of Free-Stream Turbulence on Simulations of Laminar Separation Bubbles[C] // 47th AIAA Aerospace Sciences Meeting including The New Horizons Forum and Aerospace Exposition. 2009: 1471.
192. GALBRAITH M, VISBAL M. Implicit Large Eddy Simulation of Low-Reynolds-Number Transitional Flow Past the SD7003 Airfoil[C] // 40th Fluid Dynamics Conference and Exhibit. 2010: 4737.
193. BREUER M. Effect of Inflow Turbulence on an Airfoil Flow with Laminar Separation Bubble: An LES Study[J]. Flow, Turbulence and Combustion, 2018, 101(2): 433–456.
194. CATALANO P, TOGNACCINI R. RANS Analysis of the Low-Reynolds Number Flow around the SD7003 Airfoil[J]. Aerospace Science and Technology, 2011, 15(8): 615–626.

195. COUNSIL J N N, GONI BOULAMA K. Low-Reynolds-Number Aerodynamic Performances of the NACA0012 and Selig-Donovan 7003 Airfoils[J]. Journal of Aircraft, 2013, 50(1): 204–216.
196. BERNARDOS L, RICHEZ F, GLEIZE V, et al. Prediction of Separation-Induced Transition on the SD7003 Airfoil Using Algebraic Transition Triggering[J]. AIAA Journal, 2019, 57(9): 3812–3824.
197. CORTEN G P, VELDKAMP H F. Insects Can Halve Wind-Turbine Power[J]. Nature, 2001, 412(6842): 41–42.
198. DALILIN, EDRISY A, CARRIVEAU R. A Review of Surface Engineering Issues Critical to Wind Turbine Performance[J]. Renewable and Sustainable Energy Reviews, 2009, 13(2): 428–438.
199. BONS JP. A Review of Surface Roughness Effects in Gas Turbines[J]. Journal of Turbomachinery, 2010, 132(2): 021004.
200. SAGOL E, REGGIO M, ILINCA A. Issues Concerning Roughness on Wind Turbine Blades[J]. Renewable and Sustainable Energy Reviews, 2013, 23: 514–525.
201. HAN W, KIM J, KIM B. Effects of Contamination and Erosion at the Leading Edge of Blade Tip Airfoils on the Annual Energy Production of Wind Turbines[J]. Renewable Energy, 2018, 115: 817–823.
202. BONS J P. Surface Roughness Effects[J]. Turbine Aerodynamics, Heat Transfer, Materials, and Mechanics, 2014: 613.
203. WANG M, LU X, YANG C, et al. Numerical Investigation of Distributed Roughness Effects on Separated Flow Transition over a Highly Loaded Compressor Blade[J]. Physics of Fluids, 2021, 33(11): 114104.
204. Bao N S, Huo F P, Ye Z Q, et al. Aerodynamics Performance Influence with Roughness on Wind Turbine Airfoil Surface[J]. Acta Energiae Solaris Sinica, 2005(04): 458–462.
205. Dong S C, Wang X D, Kang S. Numerical Simulationson the Effects of Surface Roughness on Airfoil Aerodynamic Performance[J]. Journal of Engineering Thermophysics, 2015, 36(06): 1238–1241.
206. SERDAR GENÇ M, KOCA K, AÇIKEL H H. Investigation of Pre-Stall Flow Control on Wind Turbine Blade Airfoil Using Roughness Element[J]. Energy, 2019, 176: 320–334.
207. LIGRANI P M, MOFFAT R J. Structure of Transitionally Rough and fully Rough Turbulent Boundary Layers[J]. Journal of Fluid Mechanics, 1986, 162: 69–98.
208. WILCOX D C. Formulation of the $k-\omega$ Turbulence Model Revisited[J]. AIAA Journal, 2008,46(11): 2823–2838.
209. PATEL V C, YOON J Y. Application of Turbulence Models to Separated Flow Over Rough Surfaces [J]. Journal of Fluids Engineering, 1995, 117(2): 234–241.
210. PATEL V C. Perspective: Flow at High Reynolds Number and Over Rough Surfaces—Achilles Heel of CFD[J]. Journal of Fluids Engineering, 1998, 120(3): 434–444.
211. HELLSTEN A, LAINE S, HELLSTEN A, et al. Extension of the -omega-SST Turbulence Model for Flows over Rough Surfaces[C] // Atmospheric Flight Mechanics Conference. 1997: 3577.
212. DURBIN P A, MEDIC G, SEO J M, et al. Rough Wall Modification of Two-Layer $k-\varepsilon$[J]. Journal of Fluids Engineering, 2001, 123(1): 16–21.
213. SPALART P R, ALLMARAS S R. A one-equation turbulence model for aerodynamic flows[R]. AIAA 92–0439, 1992.
214. SON C, KELLY M, KIM T. Boundary-Layer Transition Model for Icing Simulations of Rotating Wind Turbine Blades[J]. Renewable Energy, 2021, 167: 172–183.
215. von DEYN L H, FOROOGHI P, FROHNAPFEL B, et al. Direct Numerical Simulations of Bypass Transition over Distributed Roughness[J]. AIAA Journal, 2020, 58(2): 702–711.
216. JEONG H, SONG S J. Surface Roughness Impact on Boundary Layer Transition and Loss Mechanisms Over a Flat-Plate Under a Low-Pressure Turbine Pressure Gradient[J]. Journal of Turbomachinery, 2022, 144(1): 011005.
217. STIEGER R D. The Effect of Wakes on Separating Boundary Layers in Low Pressure Turbines[D]. [S.1.]: University of Cambridge, 2002.

218. JANISZEWSKA J M, RAMSAY R R, HOFFMANN M J, et al. Effects of Grit Roughness and Pitch Oscillations on the S814 Airfoil: TP-442–8161[R]. Golden, Colorado: National Renewable Energy Laboratory, 1996.
219. STRIPF M, SCHULZ A, WITTIG S. Surface Roughness Effects on External Heat Transfer of a HP Turbine Vane[J]. Journal of Turbomachinery, 2005, 27(1): 200–208.
220. STRIPF M, SCHULZ A, BAUER H J, et al. Extended Models for Transitional Rough Wall Boundary Layers with Heat Transfer-Part II: Model Validation and Benchmarking[J]. Journal of Turbomachinery, 2009, 131(3): 031017.
221. STRIPF M, SCHULZ A, BAUER H J. Roughness and Secondary Flow Effects on Turbine Vane External Heat Transfer[J]. Journal of Propulsion and Power, 2007, 23(2): 283–291.
222. SØRENSEN N N, BECHMANN A, ZAHLE F. 3D CFD Computations of Transitional Flows Using DES and a Correlation Based Transition Model [J]. Wind Energy, 2011, 14(1): 77–90.
223. YOU JY, KWON O J. Blending of SAS and Correlation-Based Transition Models for Flow Simulation at Supercritical Reynolds Numbers[J]. Computers & Fluids, 2013, 80: 63–70.
224. CELIK I, PATEL V. Boundary-Layer Development on Circular Cylinders[J]. Boundary-Layer Meteorology, 1982, 24(3): 281–293.
225. QIAOL, BAIJQ, HUA J, et al. Combination of DES and DDES with a Correlation Based Transition Model[C] // Applied Mechanics and Materials: Vol 444. 2014: 374–379.
226. SA JH, PARK S H, KIM C J, et al. Low-Reynolds Number Flow Computation for Eppler 387 Wing Using Hybrid DES/Transition Model[J]. Journal pf Mechanical Science and Technology, 2015, 29(5): 1837–1847.
227. GERMANO M. Properties of the Hybrid RANS/LES Filter[J] Theoretical and Computational Fluid Dynamics, 2004, 17(4): 225–231.
228. GERMANO M. On the Hybrid RANS-LES of Compressible Flows[C] // Progress in Hybrid RANS-LES Modelling. Berlin: Springer-Verlag, 2015: 253–263.
229. SÁNCHEZ-ROCHA M, MENON S. The Compressible Hybrid RANS/LES Formulation Using an Additive Operator[J]. Journal of Computational Physics, 2009, 228(6): 2037–2062.
230. HODARA J, SMITH M J. Hybrid Reynolds-Averaged Navier-Stokes/Large-Eddy Simulation Closure for Separated Transitional Flows[J]. AIAA Journal, 2017, 55(6): 1948–1958.
231. CODER J G, ORTIZ-MELENDEZ H D. Transitional Delayed Detached-Eddy Simulation of Multielement High-Lift Airfoils[J]. Journal of Aircraft, 2019, 56(4): 1302–1312.
232. XIAO Z, WANG G, YANG M, et al. Numerical Investigations of Hypersonic Transition and Massive Separation Past Orion Capsule by DDES-Tr[J]. International Journal of Heat and Mass Transfer, 2019, 137: 90–107.
233. WANG L, FU S. Modelling Flow Transition in a Hypersonic Boundary Layer with Reynolds-Averaged Navier-Stokes Approach[J]. Science in China, Series G: Physics, Mechanics and Astronomy, 2009, 52(5): 768–774.
234. KIM H J, KWON O J. Numerical Simulation of Transitional Flows Using a Blended IDDES and Correlation-Based Transition Model[J]. Computers & Fluids, 2021, 222: 104916.
235. DUAN Z, XIAO Z. Effects of External Disturbances on the Diamond-Induced Transition by Direct Numerical Simulation[J]. AIAA Journal, 2022, 60(2): 731–746.
236. SURYANARAYANAN S, GOLDSTEIN D B, BERGER A R, et al. Effect of Pressure Gradients on the Different Stages of Roughness Induced Boundary Layer Transition[J]. International Journal of Heat and Fluid Flow, 2020, 86: 108688.
237. Zhou Y L. Study of the Effect of Surface Roughness on the Hypersonic Boundary Layer Stability[D]. [S.1.]: National University of Defense Technology, 2019.
238. Zhu D H, Yuan X J, Shen Q, et al. Numerical Simulation and Mechanism Analysis of Hypersonic Roughness Induced Transition[J]. Chinese Journal of Theoretical and Applied Mechanics, 2015, 47(3): 381–388.
239. Lei Q, Jian S, Yi L, et al. Improved hybrid model for transitional separated flows over a rough compressor blade[J]. Aerospace Science and Technology, 2023, 140: 108434.
240. YIN Z, DURBIN P A. An Adaptive DES Model that Allows Wall-Resolved Eddy Simulation[J]. International Journal of Heat and Fluid Flow, 2016, 62: 499–509.

241. ACHENBACH E. Influence of Surface Roughness on the Cross-Flow around a Circular Cylinder[J]. Journal of Fluid Mechanics, 1971, 46(2): 321–335.
242. SON C, KIM T. Boundary Conditions of Flow Transition Model for Roughened Surface[J]. AIAA Journal, 2022, 60(1): 532–536.
243. HODSON HP, HOWELL R J. Bladerow Interactions, Transition, and High-Lift Aero-foils in Low-Pressure Turbines[J]. Annual Review of Fluid Mechanics, 2005, 37(1): 71–98.
244. DENTON JD. The 1993 IGTI Scholar Lecture: Loss Mechanisms in Turbomachines[J]. Journal of Turbomachinery, 1993, 115(4): 621–656.
245. WANG M, LI Z, YANG C, et al. Numerical Investigations of the Separated Transitional Flow over Compressor Blades with Different Loading Distributions[J]. Aerospace Science and Technology, 2020, 106: 106113.
246. BOESE M, FOTTNER L. Effects of Riblets on the Loss Behavior of a Highly Loaded Compressor Cascade[C] // Turbo Expo: Power for Land, Sea, and Air: Vol 3610. 2002: 743–750.
247. WANG M, LIZ, YANG C, et al. Large Eddy Simulation of the Separated Flow Transition on the Suction Surface of a High Subsonic Compressor Airfoil[J]. Physics of Fluids, 2020, 32(3): 034110.
248. DORNEY D J, LAKE J P, KING P I, et al. Experimental and numerical investigation of losses in low-pressure turbine blade rows[J]. International Journal of Turbo and Jet-Engines, 2000, 17(4): 241–254.
249. SUZEN Y, HUANG G, VOLINO R, et al. A comprehensive CFD study of transitional flows in low-pressure turbines under a wide range of operating conditions[C]. Orlando: 33rd AIAA fluid dynamics conference and exhibit, 2003.
250. Falkner V M, Skan S W. Some approximate solutions of the boundary-layer equations[J] Philosophical Magazine, 1931, 12, pp: 865-896.
251. Aziz A, Na T N. New approach to the solutions of Falkner-Skan equation[J]. AIAA Journal, 1981, 19(9), pp: 1242–1244.
252. Zhang Z X, Dong Z N, et al. Viscous Fluid Mechanics[M]. Beijing: Tsinghua University Press, 1998.
253. Cooke J C. The boundary layer of a class of infinite yawed cylinders [J]. Mathematical Proceedings of the Cambridge Philosophical Society, 1950, 46, pp: 645–648.

MIX
Papier aus verantwortungsvollen Quellen
Paper from responsible sources
FSC® C105338

If you have any concerns about our products,
you can contact us on
ProductSafety@springernature.com

In case Publisher is established outside the EU,
the EU authorized representative is:
Springer Nature Customer Service Center GmbH
Europaplatz 3, 69115 Heidelberg, Germany

Printed by Libri Plureos GmbH
in Hamburg, Germany